Lithography Process Control

TUTORIAL TEXTS SERIES

Lithography Process Control

Harry J. Levinson

Tutorial Texts in Optical Engineering
Volume TT28

Donald C. O'Shea, Series Editor
Georgia Institute of Technology

SPIE OPTICAL ENGINEERING PRESS
A Publication of SPIE—The International Society for Optical Engineering
Bellingham, Washington USA

Library of Congress Cataloging-in-Publication Data

Levinson, Harry J.
    Lithography process control / Harry J. Levinson.
      p. cm. – (Tutorial texts in optical engineering; v. TT28)
    Includes bibliographical references and index.
    ISBN 0-8194-3052-8
      1. Semiconductors—Etching. 2. Microlithography—Quality control. I. Title. II. Series.
TK7871.85.L464   1999
621.3815'2—dc21   98-44433
                                                                              CIP

Published by

SPIE—The International Society for Optical Engineering
P.O. Box 10
Bellingham, Washington 98227-0010
Phone: 360/676-3290
Fax: 360/647-1445
Email: spie@spie.org
WWW: http://www.spie.org/

Printed in the United States of America.

INTRODUCTION TO THE SERIES

The Tutorial Texts series was begun in response to requests for copies of SPIE short course notes by those who were not able to attend a course. By policy the notes are the property of the instructors and are not available for sale. Since short course notes are intended only to guide the discussion, supplement the presentation, and relieve the lecturer of generating complicated graphics on the spot, they cannot substitute for a text. As one who has evaluated many sets of course notes for possible use in this series, I have found that material unsupported by the lecture is not very useful. The notes provide more frustration than illumination.

What the Tutorial Texts series does is to fill in the gaps, establish the continuity, and clarify the arguments that can only be glimpsed in the notes. When topics are evaluated for this series, the paramount concern in determining whether to proceed with the project is whether it effectively addresses the basic concepts of the topic. Each manuscript is reviewed at the initial state when the material is in the form of notes and then later at the final draft. Always, the text is evaluated to ensure that it presents sufficient theory to build a basic understanding and then uses this understanding to give the reader a practical working knowledge of the topic. References are included as an essential part of each text for the reader requiring more in-depth study.

One advantage of the Tutorial Texts series is our ability to cover new fields as they are developing. In fields such as sensor fusion, morphological image processing, and digital compression techniques, the textbooks on these topics were limited or unavailable. Since 1989 the Tutorial Texts have provided an introduction to those seeking to understand these and other equally exciting technologies. We have expanded the series beyond topics covered by the short course program to encompass contributions from experts in their field who can write with authority and clarity at an introductory level. The emphasis is always on the tutorial nature of the text. It is my hope that over the next few years there will be as many additional titles with the quality and breadth of the first ten years.

Donald C. O'Shea
Georgia Institute of Technology

TABLE OF CONTENTS

PREFACE

In the mid-1980s the semiconductor industry underwent a fundamental change. With superior yields, reliability, and efficiency, Japanese manufacturers of integrated circuits surpassed their American competitors in terms of market share. The American semiconductor industry responded with significant quality improvement and regained the leadership market position by 1993. Today the integrated circuit business is highly competitive and global.

During the transition years I found that I was frequently consulted by lithography engineers who were attempting to apply quality improvement methods, such as statistical process control (SPC), to which they had been recently introduced. Through discussions with these engineers it became clear that there were subtle reasons why SPC could not be applied in a straightforward way to many situations which occurred in lithography. My explanations to the engineers evolved into a set of class notes, and now this Tutorial Text.

Many of the quality problems that we were trying to solve arose in conventional manufacturing situations, while others involved development pilot lines. Methods for addressing the problems that occur in the context of process development are rarely addressed in texts on process control. Another objective of this Tutorial Text is to present control methodologies applicable to development pilot lines.

To understand this text there are some prerequisites. A basic foundation in lithography science is assumed. The SPIE Handbook on *Microlithography, Micromachining and Microfabrication. Volume 1: Microlithography*[1] provides a suitable introduction. It is also assumed that the reader has had some introduction to basic statistical concepts and statistical process control. It is my intention that this text be a self-contained tutorial on lithography process control for readers familiar with the prerequisite lithography science and basic statistical process control, although some subjects may involve a higher level of mathematical sophistication than others. The text covers the subject of lithography process control at several levels. Discussions of some very basic elements of statistical process control and lithography science are included, because, when trying to control a lithography process, a number of subtle problems arise that are related to fundamental issues. To most readers, the information presented on the foundations of statistical process control should be familiar. Nevertheless, it is useful to review the foundations of statistical process control, in order to clearly identify those circumstances in which the method may be applied, and where it needs to be applied with particular care. This inclusion of basic topics also allows the text to serve as an introduction to process control for the novice lithography engineer and as a reference for experienced engineers. More advanced topics are also included to varying levels of detail. Some of these topics, such as complex processes and feedback, are discussed in considerable detail, because there is no comparable presentation available. Other topics are only introduced briefly, and the reader is referred to other texts that cover the subject quite well.

The text also contains numerous references to the extensive literature on the subject of this book. These references are intended as a guide for further study by the interested reader and are also meant to serve as an acknowledgment to the many people who have contributed over the years to improving our understanding of the lithography process and how better to control it.

A few special acknowledgments are in order. First, I want to thank my wife, Laurie Lauchlan, who tolerated the many hours I spent in the study writing this book, and who shared many of her insights on metrology. Dr. David C. Joy helpfully provided information on recent developments in the understanding of charging in low voltage SEMs. I also want to express my gratitude to Chuck DeHont, who first allowed many of these ideas first to be implemented. Finally, I want to thank the many people with whom I have worked at Advanced Micro Devices, Sierra Semiconductor, and IBM, through whose efforts the world has been improved.

Harry J. Levinson
January 1999

---

[1] *Handbook of Microlithography, Micromachining, and Microfabrication. Volume 1: Microlithography*, P. Rai-Choudhury, Ed., SPIE Press, Bellingham (1997).

Lithography Process Control

CHAPTER 1
INTRODUCTION TO THE USE OF STATISTICAL PROCESS CONTROL IN LITHOGRAPHY

Statistical methods need to be part of every lithographer's toolbox, because lithographic processes contain intrinsic levels of variation. This variation is a consequence of the nature of the world. For example, petroleum is typically the starting material from which photoresists are synthesized, and the composition of crude oil varies from well to well. Lithographic processes and tools are affected by environmental parameters such as barometric pressure and relative humidity, and these factors vary with the weather. Lithography is a manufacturing science implemented and ultimately exercised by human beings, each of whom is a unique individual, different from all others. When people are involved, there is a special element of variation interjected into the process. The analytical methods used by lithography engineers and managers must be capable of dealing with variation in equipment, materials, and people.

The objective of any process control methodology is the reduction of variation, in order to maintain conformance to standards or to meet a higher standard. Variations in gate lengths can lead to degraded yield or slower parts, which usually sell for a lower price than faster devices. Higher manufacturing costs result from variation, in the form of scrap, reduced yield, rework, and low equipment utilization. Often there are costs associated with attempts to reduce variation. Since the objective of process control is to maximize profitability, the most effective methods are those which accomplish control in the most cost-effective manner. Indeed, Dr. Walter Shewhart, the inventor of statistical process control, titled his book, *Economic Control of Quality of Manufactured Product*,[1] with the first word of the title identifying the monetary considerations motivating his methods. Generally, the most economical approaches require that particular levels of variation be tolerated, and the purpose of statistical process control is the identification of variation in excess of the norms of a controlled process.

Statistics is the mathematical science for making inferences about quantities which are probabilistic (in contrast to deterministic) in nature. Many statistical methods are therefore applicable only to situations that are random, in which events are independent of each other. A large fraction of the first three chapters of this Tutorial Text will involve the examination of commonly occurring situations in microlithography in which measurements are not independent. The discussion will lead to methods for applying statistical control techniques correctly in such situations.

A word is in order about the probabilistic versus deterministic nature of quantities that are encountered in microlithography. Nearly all lithography processes occur at levels that are well described by classical physics. For example, optical patterning involves sufficient numbers of photons for the imaging to be well characterized by classical electrodynamics and optics. Processes can be understood to a high level of accuracy, but variations occur because the choice is made, on an economic basis, to not determine actual process conditions to the level necessary to make accurate quantitative predictions. There is a significant difference between variations whose causes are unknowable versus simply unknown. Microlithography is currently far from quantum limits, where fluctuations arise from physical law.

While statistical methods are essential for identifying situations in which there are abnormal levels of variation, control is re-established only after corrective action. Effective response is achieved by using engineering science to identify and then address causes of variation. Process control is the application of engineering science, where statistical techniques are used to interpret data. This tutorial text will discuss statistics, lithography science, and the interplay between the two. Statistical methods will be the primary topic of the first three chapters, while Chapters 4 and 5 will focus on the scientific basis of image formation and pattern placement, respectively. Chapter 6 will deal with the subject of yield, with a focus on defect monitors. Process drift, feedback, active control and their relationships to statistical process control will be discussed in Chapter 7. All control methodologies rely upon data, and the integrity of the data is crucial. Metrology is difficult in the world of deep sub-micron geometries, and Chapter 8 will deal with this critical aspect of process control. In Chapter 9 we move from the mathematical and physical scientific aspects of processing to considerations of the most direct human elements, and the text is concluded with a discussion of the control of operations.

1.1 THE ASSUMPTIONS UNDERLYING STATISTICAL PROCESS CONTROL

The genius of Walter Shewhart was in identifying a methodology applicable to all random processes, regardless of their underlying probability distributions. This universal approach was achieved by exploiting the central-limit theorem, which is summarized as follows:

If individual and independent measurements are sampled from any random process in subgroups comprised of n individual measurements, and the measurements of the jth subgroup are averaged

$$\bar{x}_j = \frac{1}{n}\sum_{i=1}^{n} x_{ij} , \qquad (1.1)$$

where x_{ij} is the ith measurement in the jth subgroup, then:

1) The distribution of subgroup averages ($\bar{x}_j$) will approach a normal distribution (Gaussian or bell curve) as n gets large.

2) The distribution of subgroup averages will have the same average as the distribution of individual measurements:

$$\frac{1}{N}\sum_{j=1}^{N}\bar{x}_j = \frac{1}{nN}\sum_{i=1}^{n}\sum_{j=1}^{N}x_{ij} \xrightarrow{n,N\to\infty} \bar{x} , \qquad (1.2)$$

where N is the number of subgroups and $\bar{x}$ is the mean of the overall process.

3) The standard deviation $\sigma_{\bar{x}}$ of the distribution of averages $\bar{x}_j$ is given by:

$$\sigma_{\bar{x}} = \frac{\sigma_x}{\sqrt{n}} , \qquad (1.3)$$

where σ_x is the standard deviation of the individual measurements.

Throughout this text, a bar above a quantity indicates a mean value. Discussions of the central-limit theorem can be found in most basic texts on probability theory.[2]

According to the central-limit theorem, any random distribution can be effectively transformed to the normal distribution by subgrouping and averaging. Studies have shown that this can be accomplished effectively with subgroups as small as three or four, so long as the primary distribution does not depart too significantly from normality.[3] Moreover, the resulting normal distribution has key parameters, the average and standard deviation, that can be related to their respective values in the original random distribution. By working with the normal distributions obtained through subgrouping and averaging, conclusions about processes can be made without knowing the functional form of the original random distribution. As will be seen shortly, in lithography the choice of subgroups must be made carefully.

1.2 THE PROPERTIES OF STATISTICAL PROCESS CONTROL

Statistical process control is based upon the normal distribution, whose properties are well known. The functional form of the normal probability distribution, for a random variable x, is

$$f(x) = \frac{1}{\sigma\sqrt{2\pi}} e^{-\frac{(x-\mu)^2}{2\sigma^2}} \qquad (1.4)$$

where σ is the standard deviation of the process and μ is the process mean. This function is plotted in Fig. 1.1. Some of the well-known properties of this distribution are:

1) Data fall within ± 3σ 99.73% of the time. In other words, data fall outside of ± 3σ once every 370 times.

2) Half of the data are larger than the mean, and half are smaller. The probability of having eight points in a row above the mean is $\dfrac{1}{2^8} = \dfrac{1}{256} = 0.0039$, and the probability of having eight points below the mean is 0.0039 (0.39%).

3) Two out of three consecutive points are between 2σ and 3σ or between $-2σ$ and $-3σ$ with a frequency of 0.30%.

4) Four out of five consecutive points are between 1σ and 3σ or between $-1σ$ and $-3σ$ with a frequency of 0.54% .

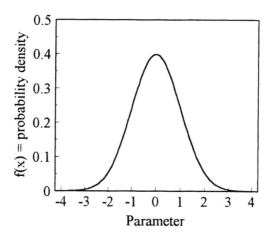

Figure 1.1. The probability density function for the standard normal distribution with a mean of zero and a standard deviation of 1.0.

These properties of the normal distribution lead to the "Western Electric Rules" of statistical process control (Fig. 1.2), so called, because SPC and these rules originated at the Western Electric Company.[4] The idea behind the Western Electric Rules is that action should be taken if events occur that are improbable based upon normal variation. For example, data ($\bar{x}_j$) should exceed ±3 $σ_{\bar{x}}$ only once every 370 times as long as the process remains in control. Suppose control charts are updated once per day. In this situation, one should find that $\bar{x}_j$ falls outside of ±3 $σ_{\bar{x}}$ slightly less than once per year, so long as the process remains in control. With such a low probability of occurrence, the existence of data outside of ±3 $σ_{\bar{x}}$ suggests that the process is no longer in control. Similarly, a long sequence of values all above or below the mean is highly improbable. For example, the probability of having eight values in a row either above or below the mean is $2 \times \left(\dfrac{1}{2} \right)^8 = \dfrac{1}{128}$. For a process in control, false alarms, indicating out-of-control conditions, occur, on average, only once in 128 values of $\bar{x}_j$. The average number of $\bar{x}_j$ values between events indicating potential out-of-control conditions, is referred to as the "average run length." This is the average number of times that $\bar{x}_j$ is determined before a Western Electric Rule is violated.

1) Any single point falls outside of the +/-3σ limits.

2) Eight successive points are above the mean, or eight successive points are below the mean.

3) Two out of three successive points are between 2σ and 3σ or between -2σ and -3σ.

4) Four out of five successive points are between 1σ and 3σ or between -1σ and -3σ.

Figure 1.2. The Western Electric Rules. Corrective action should be taken if any these situations occur.

Suppose there is a shift in the process mean. The average run length will decrease, as shown in Fig. 1.3. For no shift in the mean, the average run length is 370 when monitoring the process by the "±3σ" rule, and 128 for monitoring eight-in-a-row above or below the mean. When the mean has shifted one sigma, these values decrease to 44 and 8, for the "±3σ" rule and the "eight-in-a-row" tests, respectively. For the "eight-in-a-row" test, mean shifts are detected after the minimum number of values, eight, while the "±3σ" test is less sensitive. The process can be continued, on average, for 44 more testing intervals before an out-of-control condition is indicated, when only the "±3σ" test is used. This lack of sensitivity to small shifts in the process is a concern that will be discussed further in Section 1.5 on process capability.

Another type of process change is an increase in the standard deviation, σ. When this occurs, the "eight-in-a-row" test will indicate nothing so long as the process mean remains unchanged and the process distribution remains symmetric. On the other hand, the "±3σ" test will have a decreasing average run length with increasing σ, as shown in Fig. 1.4.

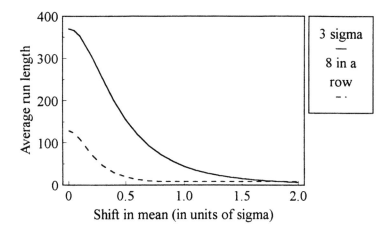

Figure 1.3. The average run length as a function of mean shift for two of the Western Electric Rules.

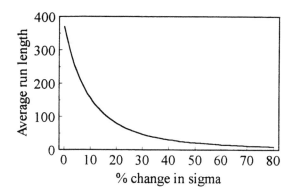

Figure 1.4. The average run length as a function of increase in the process standard deviation, for the +/-3σ rule.

1.3 SITUATIONS IN LITHOGRAPHY WHERE STATISTICAL PROCESS CONTROL CANNOT BE APPLIED NAIVELY

In order to create a process control chart, one needs to estimate $\sigma_{\bar{x}}$. This is usually accomplished in one of two ways.

1) A large number of measurements x_i can be taken, and the estimate can be calculated directly:

$$s = \sqrt{\sum_{i=1}^{M} \frac{(x_i - \bar{x})^2}{M-1}} \qquad (1.5)$$

In this equation, s is the estimate of the standard deviation for the overall process, $\bar{x}$ is the process mean, and M is the total number of measurements taken. If N subgroups of size n are taken, then

$$M = n \times N. \qquad (1.6)$$

Eq. 1.3 can then be used to calculate an estimate for $\sigma_{\bar{x}}$.

2) The calculation of the right side of Eq. (1.5), involving multiplication and square roots, is difficult without calculators or computers. Statistical process control originated in the 1920's, when such tools were unavailable. Consequently, much of the methodology of SPC is based upon estimates of standard deviations based upon ranges. The range R, which is given by:

$$R = \text{largest value - smallest value,} \qquad (1.7)$$

is easily computed for small subgroups without calculators or computers. A mean range $\bar{R}$ can be computed as well:

$$\bar{R} = \frac{1}{N} \sum_{j=1}^{N} R_j, \qquad (1.8)$$

where R_j is the range for the jth subgroup. For normally distributed measurements, the mean range of subgroup data provides an estimate for the population standard deviation according to[5]:

$$s = \frac{\bar{R}}{d_2^*}. \qquad (1.9)$$

When $\bar{R}$ has been determined from a large number of subgroups, values for the parameter d_2^* are given in Table 1.1.[6] The estimate for $\sigma_{\bar{x}}$ can then be calculated from Eq. (1.3).

The two methods described above are the most common approaches for estimating the standard deviation, which is used for determining process control limits. In one method, the standard deviation is computed directly from all data collected over time, while the other involves an estimation of the standard deviation based upon the ranges of data in the subgroups. Unfortunately, neither of these methods apply in situations which occur commonly in lithography operations. Consider a monitor of resist coatings, where a wafer is coated daily, and thickness is measured at nine locations on the wafer. From the measurements on this wafer, a daily mean and standard deviation can be computed. The results from a typical process are summarized in Table 1.2.

Subgroup size	d_2^*
2	1.128
3	1.693
4	2.059
5	2.326
6	2.534
7	2.704
8	2.847
9	2.970
10	3.078

Table 1.1. Factors for estimating the standard deviation from the mean range of subgroups.

Average uniformity across individual wafers (σ)	9 Å
Average standard deviation of wafer-to-wafer means, 10 consecutive wafers	6 Å
Standard deviation of wafer-to-wafer means over three months.	17 Å
Overall process standard deviation.	19 Å

Table 1.2. Variations observed in a monitor or resist thickness. Eq. 1.7 was used to estimate the average uniformity across individual wafers.

Suppose that the average resist thickness on the wafers provides the values for $\bar{x}_j$ that are used to generate a control chart. One might expect that the nine measurements across each wafer comprise the corresponding subgroups. The uniformity across individual wafers (first row in Table 1.2), in practice, could be obtained by determining the average range of wafer thickness and using Eq. 1.9, and control limits could be calculated using the resulting value for σ and Eq. 1.3. However, this is not the right thing to do.

For this process, the control chart of daily wafer means is charted in Fig. 1.5. Using the first row of data in Table 1.2 and Eq. 1.3 to calculate the control limits results in the absurdly tight limits about the process mean (8000 Å) of $\pm 3 \times \sigma_{\bar{x}} = \pm \dfrac{3 \times \sigma}{\sqrt{n}} = \pm 9$ Å. Here we are using the conventional Shewhart $\pm 3\sigma$ control limits about the mean that form the basis for one of the Western Electric Rules. The nine measured thicknesses across each wafer clearly do not constitute an appropriate subgroup. Control limits calculated from all of the thickness data would have been 8000 ± 3 × 19 Å = 8000 ± 57 Å about the process mean, as indicated by the data in Table 1.2. However, as will be discussed shortly, the appropriate control limits for a control chart of the daily mean resist thickness should have been determined using the data from the third row of Table 1.2, resulting in control limits of 8000 ± 3 × 17 Å = 8000 ± 51 Å. The two most common methods for estimating the process standard deviation, calculating directly from Eq. (1.5) or estimating from variations within subgroups, would have led to either inappropriately loose or excessively tight limits.

The best way to avoid using an incorrect method for calculating the standard deviation is to take a critical look at the quantity that comprises the ordinate ("y-axis") of the control chart. The control limits should be calculated from the standard deviation of that quantity. In the example just discussed, the parameter being charted was the mean thickness $\overline{x}_j$ from the single wafers that were coated daily. The resulting estimate for the control limits (about the process mean) is:

$$\pm 3s = \pm 3 \sqrt{\sum_{j=1}^{N} \frac{\left(\overline{x}_j - \overline{x}\right)^2}{N-1}} \tag{1.10}$$

The variations in the mean resist thicknesses could not be inferred from variations within wafers or from measurements taken over short periods of time (see Table 1.2). Calculations of the standard deviation using the entire set of resist thickness measurements overestimates the variation in wafer-to-wafer means, because it includes across-wafer variations.

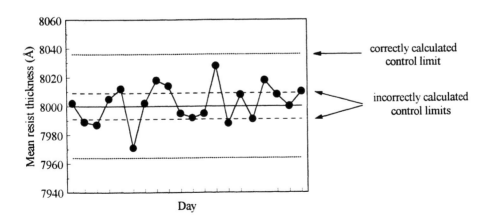

Figure 1.5. Control chart for the mean resist thickness, with incorrectly and correctly calculated control limits.

It is also important to note that the across-wafer variation and drifts in resist thickness are caused by different physical phenomena. For example, changes in barometric pressure and relative humidity will cause changes in the average resist thickness, but there will be a much smaller effect on across-wafer uniformity. Control limits based on across-wafer non-uniformity are not applicable to control charts of average resist thickness, because across-wafer non-uniformities reflect different physical phenomena than do the average thicknesses. Subgroups must be chosen carefully if variation within those subgroups are going to be used to generate control limits.

As seen from the data listed in Table 1.2, the average resist thickness varied little over the short amount of time required to coat 10 wafers, and the variation across individual wafers was less than the variation over a three month time frame. These data illustrate two types of correlations — temporal and spatial — which break randomness. Wafers coated close in time to each other will tend to have similar resist thicknesses. Across individual wafers, the resist thickness will not vary rapidly over short distances. These situations, in which measurements are related non-randomly to the values of other parameters, will be discussed in Section 2.3.

1.4 NON-NORMAL DISTRIBUTIONS

In the preceding example, the subgroup size = 1. A single wafer was coated daily, and it was the mean thicknesses of the resist on the individual wafers that were subjected to process control. As described in Section 1.1, the methodology of SPC is based upon the ability to transform all random distributions to ones that are normally distributed, by means of subgroup averaging and the central-limit theorem. When subgroup sizes of one are used, care must be taken to ensure that the process being monitored varies normally, or the conventional SPC formalism may lead to incorrect conclusions.

There are a number of circumstances that arise in lithography operations in which it is inconvenient to generate subgroups of size greater than one. The example of resist coating monitors, discussed in the previous section, was one example. There are many reasons for restricting subgroups to the minimum size, which is one:

- High measuring costs. Metrology equipment, for measuring defects, film thickness, linewidths and overlay, is expensive. Reducing the number of measurements reduces the expenditures required for measurement equipment.
- The need to maximize equipment utilization. Processing wafers for the purpose of monitoring equipment detracts from productive time, in which processing equipment is used to make saleable product. It is clearly desirable to minimize the time used for processing material that cannot be sold.
- Data collection is very time consuming. The amount of time required for people to process, measure, and analyze equipment monitors tends to

increase with the number of wafers processed. Minimizing the number of wafers reduces the level of human resources needed to control the process. For all of these reasons, it is frequently desirable in lithography operations to have sub-groups of size equal to one. Since conventional SPC formalism can be conveniently applied to individual measurements in situations where the processes do vary normally, and may be problematic where the processes do not, it is useful to have methods to determine whether processes are distributed normally or not. There are graphical and analytical methods that can test for normality. The former involves the use of *normal probability paper*, which is derived from the graph of the cumulative probability of the normal distribution, shown in Fig. 1.6. The cumulative probability is obtained by integrating Eq. 1.4:

$$\text{Cumulative probability} = \int_{-\infty}^{x} \frac{1}{\sigma\sqrt{2\pi}} e^{-\frac{(x'-\mu)^2}{2\sigma^2}} dx' \qquad (1.11)$$

Normal probability paper is created by stretching Fig. 1.6 along the ordinate nonlinearly, such that the cumulative probability curve turns into a straight line. The result is shown in Fig. 1.7.

To use normal probability paper, do the following:
1) Rank the data from smallest to largest, $i = 1, ..., n$.
2) Determine the percentage value, F_i, for the ranked data.

$$F_i = \frac{100(i - 0.5)}{n} \quad 1 = 1, ..., n. \qquad (1.12)$$

3) Plot the data on normal probability paper, where the data values are plotted on the (linear) abscissa and the percentage value is plotted on the ordinate.

Data that are normally distributed will plot as nearly a straight line. A convenient source of normal probability paper is the book, *Graph Paper from Your Copier*,[7] from which the purchaser is free to make photocopies. Alternatively, there are a number of software packages that plot the data onto normal probability scales automatically. A listing of software packages which have this capability can be found in the annual software directory in the magazines, *Quality Progress* (usually in the April issue) or *Quality* (December issue).

Example: Consider the following 25 ranked measurements:
3.8, 4.6, 4.6, 4.9, 5.2, 5.3, 5.3, 5.4, 5.6, 5.6, 5.7, 5.8, 5.9, 6.0, 6.0, 6.1, 6.3, 6.3, 6.4, 6.5, 6.6, 6.8, 7.0, 7.4, 7.6.

These data are plotted on normal probability paper and shown in Fig. 1.8. The straight line in Fig. 1.8 is for the normal distribution with the same mean and standard deviation as these data.

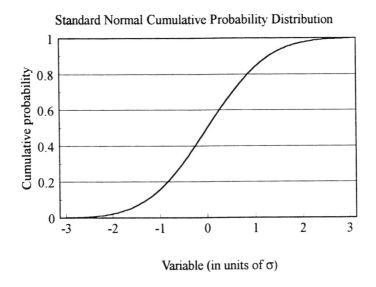

Figure 1.6. The cumulative normal distribution.

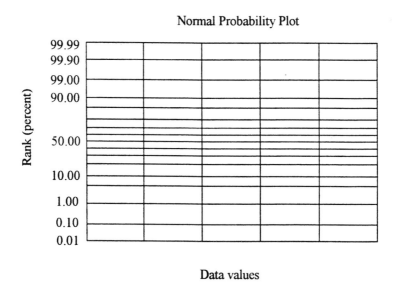

Figure 1.7. Normal probability paper.

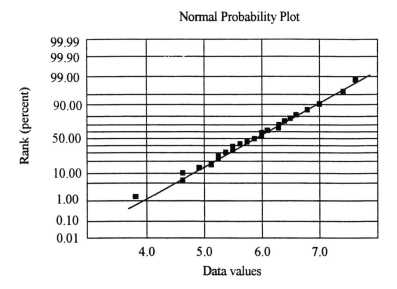

Figure 1.8. Normal probability plot for the data discussed in the text.

Normal probability paper was a tool from an era in which computations were difficult and laborious. The primary deficiency of this method is the degree of subjectivity in deciding whether data fit a straight line well or not. There are numerous analytical and objective methods for testing normality, the most common of which is the χ^2 (chi-squared) test,[8] which is based upon evaluating the statistic:

$$\chi^2 = \frac{(O_1 - E_1)^2}{E_1} + \frac{(O_2 - E_2)^2}{E_2} + \cdots + \frac{(O_2 - E_k)^2}{E_k}, \qquad (1.13)$$

where O_i is the observed frequency (number of occurences) for event or category i, and E_i is the expected frequency based upon the assumption of normality. For example, an event or category might consist of having values within a particular interval. Even for a normal distribution, χ^2 usually $\neq 0$, because of random variation. A test is needed to determine when the value of χ^2 is larger than might be expected on the basis of typical random variation. When this occurs, the assumption of normality can be questioned.

The χ^2 test involves several steps:

1) From the data, calculate the estimated mean $(\bar{x})$ and standard deviation (s), which will be used to determine the expected frequencies from a normal distribution where $\mu = \bar{x}$ and $\sigma = s$.

2) With the parameters estimated in the prior step, it is possible to compute the expected frequency of data that occur between any two values. The range of the data should be completely divided into categories such that the expected

frequency is at least five in each category. The requirement that there are at least five in each category ensures robustness of the test in situations where the actual distribution deviates significantly from normality.

3) With the range of data divided into categories, the actual occurrences O_i within each category are computed.

4) Compute χ^2 using Eq. 1.13.

5) Select a desired Type I error level, α. This is the probability that the hypothesis of normality will be rejected even though the distribution truly is normal. Typical values for α are 0.05 to 0.10. The larger α is, the more stringent the testing.

6) Look up $\chi^2_{1-\alpha,\nu}$ in a table of the chi-square distribution or calculate it. Tables for the chi-square distribution can be found in most texts on statistics. The degree of freedom is $\nu = k - 3$, where k is the number of categories.

7) If $\chi^2 < \chi^2_{1-\alpha,\nu}$, then the distribution can be assumed to be normally distributed to the Type I error level of α.

The data in example above can be used to exemplify the use of the χ^2 test. The mean and standard deviation of the data set are:

$$\bar{x} = 5.868 \tag{1.14}$$

$$s = 0.892 \tag{1.15}$$

With these parameters, one can calculate the expected frequencies. It is necessary to create categories in which the expected frequency is at least 5. Additionally, it should be noted that the best test has the most comparisons between expected and actual frequency, i.e., when there are the most categories. For the data in the above example, one has the largest number of categories, with at least five events in each category, when there are five categories with an expected frequency of five in each. The probability of being in any particular category is 0.20. The boundary values that lead to this division of the data are listed in Table 1.3. These were obtained as follows. The first category consists of the smallest values. The lower boundary for such a category is $-\infty$ and the upper boundary u is given by the normal probability distribution with mean and standard deviation given in Eqs. 1.14 and 1.15:

$$\int_{-\infty}^{u} \frac{1}{s\sqrt{2\pi}} e^{-\frac{(x'-\bar{x})^2}{2s^2}} dx' = \frac{5}{25} = 0.2. \tag{1.16}$$

The boundary u forms the lower boundary for the next category, and its upper boundary is calculated similarly.

Continuing with the rest of the data, one obtains the results shown in Table 1.3, from which χ^2 is calculated to be 0.80. Letting $\alpha = 0.05$, we find that

$$\chi^2_{1-\alpha,\nu} = \chi^2_{0.95,2} = 5.99. \tag{1.17}$$

In this case,

$$\chi^2 = 0.80 < 5.99 = \chi^2_{1-\alpha,\nu}, \tag{1.18}$$

and we can conclude that the data are consistent with the assumption of a normal distribution.

Boundary value	Original data	Expected frequency	Actual frequency
$-\infty$ to 5.12	3.8, 4.6, 4.6, 4.9	5	4
5.12 to 5.65	5.2, 5.3, 5.3, 5.4, 5.6, 5.6	5	6
5.65 to 6.10	5.7, 5.8, 5.9, 6.0, 6.0	5	5
6.10 to 6.62	6.1, 6.3, 6.3, 6.4, 6.5, 6.6	5	6
6.62 to $+\infty$	6.8, 7.0, 7.4, 7.6	5	4

Table 1.3. Summary of the parameters used for applying the χ^2 test to the data from the above example.

The χ^2 test indicates whether measurements from a process are distributed normally or not. There are other tests for non-normality that involve measures of the shape of the distribution,[9] and which provide additional information that is often useful to lithographers. Two characteristics, symmetry and peakedness, are of particular interest, and each of these can be defined rigorously, i.e., mathematically.

First, a preliminary term needs to be defined. Suppose there are N measurements from a process. Then, the rth moment about the mean, m_r,

$$m_r = \frac{\sum_{i=1}^{N}(x_i - \bar{x})^r}{N}, \tag{1.19}$$

where $r = 1, 2, 3, \ldots$ Note that the standard deviation is the square root of the second moment about the mean:

$$s = \sqrt{m_2}. \tag{1.20}$$

Skewness is the term that mathematicians use to refer to the symmetry of a distribution. The coefficient of skewness is defined as

$$a_3 = \frac{m_3}{\left(\sqrt{m_2}\right)^3}. \tag{1.21}$$

Note that this is a dimensionless quantity. Symmetric distributions, such as the normal distribution, have skewness $= a_3 = 0$. A distribution with positive skewness is shown in Fig. 1.9.

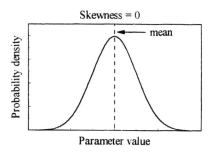 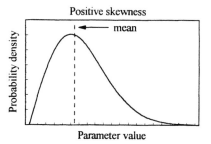

Figure 1.9. Examples of skewness.

Another shape parameter is the degree of peakedness, called kurtosis, defined by:

$$a_4 = \frac{m_4}{m_2^2}. \tag{1.22}$$

Kurtosis is also dimensionless. For a normal distribution, $a_4 = 3$. Distributions for which $a_4 = 3$ are also know as mesokurtic. For distributions that have flatter peaks than the normal distribution's bell curve (see Fig. 1.10):

$$a_4 < 3 : \text{platykurtic} \tag{1.23}$$

Distributions may also have sharper peaks than a bell curve:

$$a_4 > 3 : \text{leptokurtic} \tag{1.24}$$

Platykurtic distributions arise frequently in the measurement of overlay, which will be discussed in Chapter 5.

The shape of the distribution can often reveal some of the nature of underlying sources of variation. Consider a situation in which defocus is a significant source of linewidth variation (Fig. 1.11). In many situations the linewidth will vary in only one direction as the focus is changed. For the data shown in Fig. 1.11, the linewidth always decreases as focus is moved away from best focus. (This behavior — decreasing linewidths because of defocus —

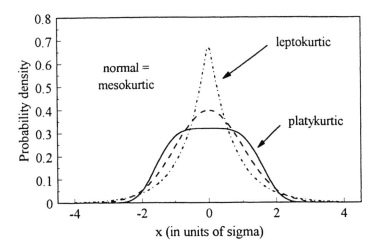

Figure 1.10. Examples of distributions with different types of kurtosis.

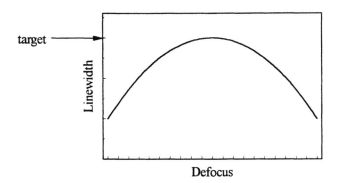

Figure 1.11. As a consequence of defocus, the linewidth always becomes smaller than the target value, leading to a skewed distribution.

is typical for isolated chrome lines printed into positive photoresist. Behavior can vary with pitch and the tone of the photoresist.[10] In this text, a line refers to a long and narrow piece of photoresist, while a space refers to a long and narrow trench in a sheet of resist.) In such a situation, the effect of defocus will be to shift the linewidths in only one direction, thus creating a skewed distribution. For example, suppose the focus varies normally, with a mean of zero and a standard deviation of σ, and if the linewidth w varies with defocus z according to

$$w = az^2, \qquad (1.25)$$

then the linewidth varies as a consequence of the variation in focus[3]:

$$\text{probability density} = \frac{1}{\sigma\sqrt{2\pi aw}}\,e^{-\frac{w}{2a\sigma^2}} \qquad \text{if } w > 0 \qquad (1.26)$$

$$= 0 \quad \text{if } w \le 0 . \qquad (1.27)$$

This probability density function is shown in Fig. 1.12.

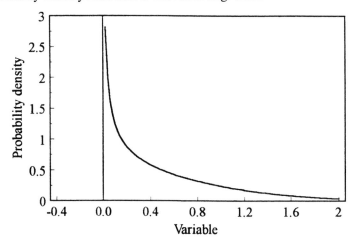

Figure 1.12. Probability density of Eq. 1.26. The ordinate is in units of σ^2/a.

Knowing that some sources of variation produce skewed distributions, one can use skewness as a diagnostic tool. Significant skew in the distribution of linewidths, in a particular direction, can be the result of poor focus control. The appearance of skew in linewidth distributions might be a clue that there is a problem with defocus, rather than some other source of increased linewidth variation. Given the number of possible sources of increased process variation, information that can reduce the number of potential factors is invaluable for expeditiously returning a process to a controlled state.

Finite samples from a normally distributed process will produce non-zero values for skewness and values for kurtosis which do not equal three, because of random variation. Levels of statistical significance must be established in order to distinguish between real deviations from normality and natural statistical variation, i.e., by how much can we expect skewness to deviate from zero, or kurtosis to deviate from three even when the process is normally distributed, but the sample size is small. The expected ranges of skewness and kurtosis for a normally distributed process are given in Tables 1.4 and 1.5. Values for skewness within the confidence intervals shown in Table 1.4, and values of kurtosis within the confidence intervals shown in Table 1.5, can be expected from normal statistical variations.

Example. The skewness for the distribution of Eqs. 1.26 and 1.27 can be calculated. Since this is a continuous distribution, integrals are required to calculate the moments. If p(w) is the probability distribution for the linewidth w, then the rth moment about the mean μ is given by:

$$m_r = \int p(w)(w - \mu)^r dw \qquad (1.28)$$

Evaluating the integrals, one finds that $\mu = a\sigma^2$, $m_2 = 2a^2\sigma^4$, $m_3 = 8a^3\sigma^6$, and consequently, $a_3 = \sqrt{2}^3 \approx 2.82$. Looking at Table 1.4, one would expect to be able to distinguish this distribution from a normal distribution with a relatively small sample size.

Sample size	95% confidence interval	99% confidence interval
25	± 0.714	± 1.073
30	± 0.664	± 0.985
35	± 0.624	± 0.932
40	± 0.587	± 0.869
45	± 0.558	± 0.825
50	± 0.533	± 0.787
60	± 0.492	± 0.723
70	± 0.459	± 0. 673
80	± 0.432	± 0.631
90	± 0.409	± 0.596
100	± 0.389	± 0.567

Table 1.4. The 95% and 99% confidence intervals for skewness, assuming a normal universe.[6]

Sample size	95% confidence interval		99% confidence interval	
	Lower interval boundary	Upper interval boundary	Lower interval boundary	Upper interval boundary
25	1.91	4.16	1.72	5.30
30	1.98	4.11	1.79	5.21
40	2.07	4.06	1.89	5.04
50	2.15	3.99	1.95	4.88
75	2.27	3.87	2.08	4.59
100	2.35	3.77	2.18	4.38

Table 1.5. The 95% and 99% confidence intervals for kurtosis, assuming a normal universe.[6]

1.5 PROCESS CAPABILITY

Statistical process control provides a method for identifying changes in processes, by comparing current processes to prior levels of performance. The requirements of a process, in terms of what is needed to produce good products, do not enter into the SPC methodology. A process may be very stable, but unable to perform within specifications. A process is said to be capable if it is able to meet the requirements of the technology (Fig. 1.13). Capability indices are metrics for gauging process performance versus requirements. The most fundamental capability index is C_p, which measures the width of the process distribution, relative to the specification limits[11]

$$C_p = \frac{USL - LSL}{6\sigma},$$ (1.29)

where USL is the upper specification limit and LSL is the lower specification limit. Note that C_p is a dimensionless number. Large values for C_p indicate a tight distribution, relative to requirements. Hence, bigger is better.

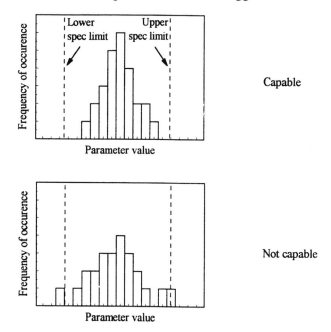

Figure 1.13. Histograms of data from a capable process and from a process that is not capable.

The two processes in Fig. 1.14 have the same C_p, even though the process centered in the middle of the specification limits is clearly more desirable. In order to incorporate process centering into the measure of process capability, additional indices have been defined. The most commonly used metric is C_{pk}[10]:

$$C_{pk} = (1-k)C_p, \qquad (1.30)$$

where

$$k = \frac{2|T - \mu|}{USL - LSL}, \qquad (1.31)$$

T is the process target, and μ is the process mean. When the process mean equals the process target, $C_p = C_{pk}$. As the process mean moves away from the target value, C_{pk} decreases.

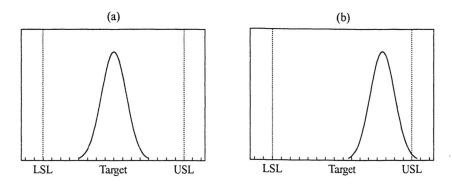

Figure 1.14. Two processes with the same $C_p = 2.0$.

There are several disadvantages to C_{pk}. First, the quantity is not an analytical function in its parameters because of the absolute value in Eq. 1.31 (the derivatives $\dfrac{\partial}{\partial T}$ and $\dfrac{\partial}{\partial \mu}$ do not exist when $T = \mu$). Second, it does not weigh shifts in the process mean heavily, compared to increases in the width of the distribution. To address these two deficiencies, another capability index has been defined:[12, 13]

$$C_{pm} = \frac{C_p}{\sqrt{1 + \frac{(T - \mu)^2}{\sigma^2}}} \qquad (1.32)$$

For the two processes in Fig 1.14, we have the process capability values shown in Table 1.6. Note that the metric C_{pm} is smaller than C_{pk} for the same process, when there is a mean shift, since the C_{pm} index emphasizes shifts of the process mean.

	Process A	Process B
C_p	2.0	2.0
C_{pk}	2.0	0.85
C_{pm}	2.0	0.56

Table 1.6. Process capability indices for the processes shown in Fig. 1.13.

Process capability is a very powerful management and engineering tool. With a single number, one is able to assess the extent to which a process can meet the needs of the product. During development, processes tend to have low values for C_{pm}, quite often < 1.0. As the processes mature, C_{pm} increases, as a consequence of process learning. However, in the ever-changing microelectronics industry, processes are usually rendered obsolete before manufacturing engineers have sufficient opportunity to bring them to full maturity. Process capability can be made large by having wide specification limits, but in the semiconductor electronics business this often results in uncompetitive products. Values for C_{pm} in excess of three, often experienced in other industries,[14] are almost never achieved in the world of microlithography. Typical values for C_{pm} in microlithography manufacturing are given in the following table:

$2.0 < C_{pm}$	World class
$1.3 < C_{pm} < 2.0$	OK
$1.0 < C_{pm} < 1.3$	Significant improvement is needed
$C_{pm} < 1.0$	In serious trouble

Table 1.7. Interpretation of process capability.

Process capability forms the basis for the "Six-sigma" methodology advocated by Motorola.[15] Suppose one has a normally distributed process with $C_p = 2.0$. When the process is centered ($C_p = C_{pm}$), only two parts per billion will be out of specifications. Suppose the process mean shifts by 1σ. In this situation, only 0.3 parts per million will fall out of specifications, which still represents good performance. Since SPC methodology cannot identify process shifts immediately after they occur, a process where the upper and lower specification limits are 6σ from the target, prior to a shift in the process, will continue to produce quality products between the time a shift occurs and the time of detection and corrective action. Such "Six-sigma" processes ensure product quality. On the other hand, suppose one has a process where $C_p = 1.33$, which is a "4σ" process. As long as the process mean remains on target, only 63 parts per million will fall out of specifications, which is adequate for many microelectronic devices. However, with a shift in the mean of 1σ, 1350 parts per million will fall out of specifications, and this may no longer meet product requirements. Note that the relationship between the standard deviation and the fraction of non-conforming measurements given in this section holds only for normally distributed processes. For non-normal processes, the assessment summarized in Table 1.7 would probably require adjustment, since there would be different relationships between the values of the process capability indices and the fraction of products which are non-conforming.

Process capability indices must be interpreted carefully. Suppose that a device will function only if all linewidths are within specifications. If the process for producing this device is centered to the target linewidth, normally distributed, and $C_p = 1.0$, there will be 0.3 parts per million out of specifications. If the linewidths that are outside of specifications are distributed randomly across the wafers, the process would be virtually incapable of yielding 256M or 1G DRAMs. Fortunately, linewidths do not vary randomly across wafers, so processes with $C_p = 1.0$ are capable of producing such products. The non-random distribution of linewidths and overlay will be discussed in subsequent chapters.

Process capability indices are based upon the process standard deviation and mean and will have uncertainties associated with knowing each of these parameters only within the range of a confidence interval.[16] Interpretations of process capability should be made with this in mind. With limited data there are opportunities for unwarranted concern or optimism.

---

[1] W. A. Shewhart, *Economic Control of Quality of Manufactured Product*, D. van Nostrand Co., New York, (1931).

[2] H. J. Larson, *Introduction to Probability Theory and Statistical Inference*, John Wiley & Sons (1969).

[3] D. C. Montgomery, *Introduction to Statistical Quality Control*, John Wiley & Sons, New York (1996).

[4] Western Electric's *Statistical Quality Control Handbook*, Delmar Printing Company, Charlotte (1985).

[5] A. J. Duncan, *Quality Control and Industrial Statistics*, 5th Edition, Irwin, Homewood, Illinois (1986).

[6] M. G. Natrella, *Experimental Statistics*, U.S. Government Printing Office, Washington, D. C. (1966).

[7] J. S. Craver, *Graph Paper from Your Copier*, HPBooks, Tucson, AZ (1980).

[8] S. S. Shapiro, *How to Test Normality and Other Distributional Assumptions*, American Society for Quality Control, Milwaukee (1986).

[9] P. P. Ramsey and P. H. Ramsey, "Simple Tests of Normality in Small Samples," *Journal of Quality Technology*, pp. 299 – 313 (1990).

[10] M. Fujimoto, T. Hashimoto, T. Uchiyama, S. Matsuura, and K. Kasama, "Comparison Between Optical Proximity Effect of Positive and Negative Tone Patterns in KrF Lithography," SPIE Vol. 3051, pp. 739 – 750 (1997)

[11] V. E. Kane, "Process Capability Indices," *Journal of Quality Technology*, Vol. 18, pp. 41 – 52 (1986).

[12] L. K. Chan, S. W. Cheng, and F. A. Spiring, "A New Measure of Process Capability: C_{pm}," *Journal of Quality Technology*, Vol. 20, pp. 162 – 175 (1988).

[13] R. A. Boyles, "The Taguchi Capability Index," *Journal of Quality Technology*, Vol. 23, pp. 17 – 26 (1991).

[14] D. J. Wheeler, *A Japanese Control Chart*, SPC Press, Knoxville (1986).

[15] F. R. McFadden, "Six-Sigma Quality Programs," *Quality Progress*, pp. 37 – 42 (June, 1993).

[16] Y-M. Chou, D. B. Owen, and S. A. Borrego, "Lower Confidence Limits on Process Capability Indices," *Journal of Quality Technology*, Vol. 22(3), pp. 223 – 229 (1990).

CHAPTER 2
SAMPLING

....

2.1 CHOOSING THE PROPER SAMPLE SIZE

An engineer responsible for resist processing had to address the following problem. His resist supplier provided materials with batch-to-batch photospeed controlled to ± 3%. For his process, this corresponded to linewidth changes of ± 8 nm about the target value of 250 nm. The engineer wanted to determine the photospeed of a new batch of resist. This information would be used to adjust exposure doses in order to continue to pattern production wafers with linewidths at their target values. The engineer wanted to expose the minimum number of wafers in order to guarantee that the linewidths would change, on average, by no more than ± 1.5 nm after the facility changed to the new batch of resist. Because the process had been running production volumes for several months, the engineer knew that the process with the old batch of resist was well centered on the target, that is, the process mean

$$\mu_{old} = 250 \text{ nm.} \tag{2.1}$$

The engineer also knew the (short term) wafer-to-wafer variation very well:

$$\sigma = 2.0 \text{ nm.} \tag{2.2}$$

Keeping the exposure dose fixed, it was reasonable to assume that σ would be unchanged with the new batch of resist, only that μ would change.

Because of the wafer-to-wafer variation, estimates of the new process mean (μ_{new}) could be expected to vary, introducing uncertainty into the determination of μ_{new}. Such uncertainty is a function of sample size. Suppose that n wafers were used to estimate μ_{new}. Because of the random wafer-to-wafer variation, the resulting sample mean ($\bar{x}$) would not equal μ_{new} exactly. Processing and measuring groups of n wafers repeatedly would produce values for $\bar{x}$ which would vary about μ_{new} with a standard deviation of

$$\sigma_{\bar{x}} = \frac{\sigma}{\sqrt{n}}. \tag{2.3}$$

This result is true regardless of the nature of the parent distribution; it does not need to be normal. From the central-limit theorem, it is understood that repeated measurements of the sample mean $\bar{x}$ produce a distribution that is approximately normal, regardless of the parent distribution.

Suppose the resist engineer wanted to know the new process mean to within ± 1.5 nm to a 95% level of confidence or better. Then

$$1.96 \, \sigma_{\bar{x}} \leq 1.5 \, \text{nm}, \tag{2.4}$$

because 95% of all data fall with ± 1.96 σ of the mean for a normal distribution. That is,

$$1.96 = z_{1-\frac{\alpha}{2}} \tag{2.5}$$

where $\alpha = 0.05$ and $z_{1-\frac{\alpha}{2}}$ is the value of $\frac{x-\mu}{\sigma}$ such that $\frac{\alpha}{2}$ is the area of the upper tail of the standard normal distribution. (See Eq. 1.4 and Fig. 2.1.) Using Eqs. 2.3 and 2.4 and the known value for σ,

$$\sqrt{n} \geq 2.61, \tag{2.6}$$

or

$$n \geq 6.8 \,. \tag{2.7}$$

Thus, the engineer needed at least seven wafers to determine the dose for the new batch of photoresist to the desired level of confidence, so long as the process was stable, and the old process was expected to produce mean linewidths of 250 nm.

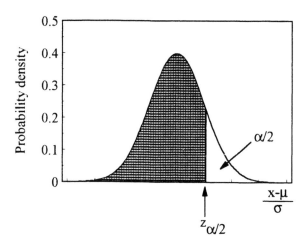

Figure 2.1. The standard normal distribution (mean = 0 and standard deviation = 1), where α/2 is the area under the upper tail.

If there was significant long-term drift to the process, then the engineer would need to expose wafers with both the old and new resist at nearly the same time, and compare means. By collecting data from the old and new resists together, equipment drifts would factor out of the comparison. Since the engineer was interested in knowing the difference between the two batches of photoresist, he would determine

$$\bar{x}_{old} - \bar{x}_{new}, \tag{2.8}$$

where $\bar{x}_{old}$ is the mean linewidth measured on wafers processed with the old photoresist, and $\bar{x}_{new}$ is the mean measured linewidth using the new batch of photoresist. Again, because of statistical variations, these measured quantities ($\bar{x}_{old}, \bar{x}_{new}$) will not exactly equal the actual process means (μ_{old}, μ_{new}). The $100 \times (1 - \alpha)$ % level of confidence for $\mu_{old} - \mu_{new}$ is given by[7]

$$(\bar{x}_{old} - \bar{x}_{new}) - z_{1-\frac{\alpha}{2}}\sqrt{\frac{\sigma_{old}^2}{n_{old}} + \frac{\sigma_{new}^2}{n_{new}}} \leq \mu_{old} - \mu_{new} \leq (\bar{x}_{old} - \bar{x}_{new}) + z_{1-\frac{\alpha}{2}}\sqrt{\frac{\sigma_{old}^2}{n_{old}} + \frac{\sigma_{new}^2}{n_{new}}}, \tag{2.9}$$

where σ_{old} and σ_{new} are the standard deviations of wafers processed with the old and new resists, and n_{old} and n_{new} are the number of wafers processed using the old and new resists, respectively. From Eq. 2.9 one can calculate the number of wafers required for $\bar{x}_{old} - \bar{x}_{new}$ to provide a sufficiently accurate estimation of $\mu_{old} - \mu_{new}$.

Situations exist in which the standard deviation is unknown. In such situations, the t distribution is used, instead of the normal distribution. If the standard deviations for the two processes are approximately the same, then the $100 \times (1 - \alpha)$ % level of confidence for $\mu_{old} - \mu_{new}$ is given by[7]

$$(\bar{x}_{old} - \bar{x}_{new}) - t_{1-\frac{\alpha}{2},v} s_p \sqrt{\frac{1}{n_{old}} + \frac{1}{n_{new}}} \leq \mu_{old} - \mu_{new} \leq (\bar{x}_{old} - \bar{x}_{new}) + t_{1-\frac{\alpha}{2},v} s_p \sqrt{\frac{1}{n_{old}} + \frac{1}{n_{new}}} \tag{2.10}$$

where

$$s_p^2 = \frac{(n_{old} - 1)s_{old}^2 + (n_{new} - 1)s_{new}^2}{n_{old} + n_{new} - 2}, \tag{2.11}$$

and the number of degrees of freedom v is given by:

$$\nu = n_{old} + n_{new} - 2, \tag{2.12}$$

and s_{old} and s_{new} are the sample standard deviations for the old and new processes, respectively. The factor $t_{1-\frac{\alpha}{2},\nu}$ is the analog of $z_{1-\frac{\alpha}{2}}$ for the t-distribution with ν degrees for freedom, where $\frac{\alpha}{2}$ is the area of the upper tail.

The preceding discussion focused on the number of measurements required to determine the mean to a sufficient level of confidence. For the sample standard deviation there are no relationships analogous to Eq. 2.3, which are valid regardless of how the data are distributed. The variations in the standard deviation are known for the normal distribution. If x is normally distributed with variance σ^2 and

$$s^2 = \frac{\sum_{i=1}^{n}(x_i - \bar{x})^2}{n-1} \tag{2.13}$$

is the sample variance based upon a random sample of size n, then $\frac{(n-1)s^2}{\sigma^2}$ has a χ^2 distribution with $\nu = n-1$ degrees of freedom. The χ^2 probability distribution for the random variable x is given by:

$$f(x) = \frac{x^{\frac{\nu}{2}-1} e^{-\frac{x}{2}}}{2^{\frac{\nu}{2}} \Gamma\left(\frac{\nu}{2}\right)}. \tag{2.14}$$

Confidence intervals at the $100 \times (1 - \alpha)\%$ level are found from[4]

$$1 - \alpha = \text{Prob.}\left(\chi^2_{1-\alpha/2} \leq \frac{(n-1)s^2}{\sigma^2} \leq \chi^2_{\alpha/2} \right), \tag{2.15}$$

where χ^2_β is the point on the χ^2 distribution illustrated in Fig. 2.2. From Eq. 2.15, confidence intervals for σ can be constructed:

$$B_L s < \sigma < B_U s, \tag{2.16}$$

where

$$B_L = \sqrt{\frac{n-1}{\chi_{\alpha/2}^2}} \qquad (2.17)$$

and

$$B_U = \sqrt{\frac{n-1}{\chi_{1-\alpha/2}^2}} \qquad (2.18)$$

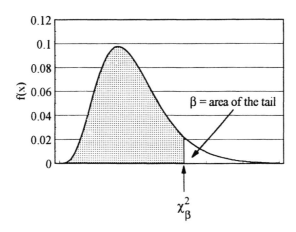

Figure 2.2. The chi-square distribution.

Values for χ_β^2 can be obtained from tables or computed directly to obtain the desired confidence interval.

Suppose we sample a process for which neither the mean nor the standard deviation are known and find

$$\bar{x} = 250 \text{ nm} \qquad (2.19)$$

and

$$3s = 25 \text{ nm}. \qquad (2.20)$$

The 95% confidence intervals for the mean and standard deviation are given in Table 2.1. Because the χ^2 distribution is asymmetric, and it is a distribution for the variance, the confidence interval for the standard deviation is not symmetric. Note that the uncertainty in the standard deviation is greater than the uncertainty of the mean.

95% confidence intervals

	Mean	Standard deviation
10 measurements	$\overline{X} - 6.3 < \mu < \overline{X} + 6.3$	$17.2 < 3\sigma < 45.6$
25 measurements	$\overline{X} - 4.8 < \mu < \overline{X} + 4.8$	$19.5 < 3\sigma < 34.8$
100 measurements	$\overline{X} - 1.7 < \mu < \overline{X} + 1.7$	$22.0 < 3\sigma < 29.0$

Table 2.1. Confidence intervals for the mean and standard deviation from the sample statistics of Eqs. 2.19 and 2.20.

These uncertainties are significant when establishing control charts for the first time, which requires that the process mean and the $\pm 3\sigma$ limits be determined. Errors in determining the mean and standard deviation will result in control charts that either fail to signal out-of-control conditions or produce numerous false alarms. While the sample standard deviation was given by Eq. 2.20, the actual value for 3σ could have been as high as 45.6 or as small as 17.2, to the 95% level of confidence, if there had been only ten measurements. Control charts established with the value for the standard deviation calculated from the first 10 measurements could have limits much too loose or too tight. The number of measurements needed to establish the control limits is driven primarily by the uncertainty in the standard deviation. In terms of correctly determining control limits, the uncertainty in the standard deviation is more significant than the uncertainty in the mean. However, for the very reason that the "eight-in-a-row" test is a sensitive indicator of a shift in the process mean, small errors in estimates of process means will result in control charts that indicate out-of-control conditions for the "eight-in-a-row" test.

The proper method for establishing control charts for newly established processes has been the subject of discussion in the quality control literature.[1, 2] Frequently it is recommended that control charts should not be implemented until they contain at least 25 data points. As one can see from Table 2.1, there will be considerable uncertainty in the control limits even for control charts with 25 entries, and out-of-control indications by the Western Electric Rules should be interpreted with this in mind. Incorrect control limits can lead to excessive numbers of false alarms or insensitivity to real process shifts, either of which is possible when control charts are set up using small amounts of data.

A final problem in which sample size is an important consideration involves estimates of variation from nested components. Consider, for example, the data in Table 1.2. The resist thickness across individual wafers varies with $\sigma = 9$ Å. If this thickness variation is random, then there will be an apparent wafer-to-wafer thickness variation, as a consequence of Eq. 2.3, for small values of n. That is, the across-wafer variation leads to an apparent wafer-to-wafer variation due to statistical fluctuations inherent in small sample sizes. This effect is particularly pronounced when across-wafer variations are large compared to wafer-to-wafer or lot-to-lot variations. When calculating the total process variation from nested components, the effects of small sample sizes must be taken into account. If we consider the total variance to be given by:

$$\sigma^2_{total} = \sigma^2_{within-wafer} + \sigma^2_{wafer-to-wafer} + \sigma^2_{lot-to-lot} \qquad (2.21)$$

then the total variation can be overestimated because sample estimates of the wafer-to-wafer and the lot-to-lot variations will contain contributions from the within-wafer variations. The wafer-to-wafer variation arises from the difference in the average resist thickness from one wafer to another. Because of Eq. 2.3, there will be a measured wafer-to-wafer variation, even if the actual averages of resist thickness change little from one wafer to another.

2.2 MEASUREMENT LOCATION CONSIDERATIONS

Consider a stepper with a magnification error. If overlay is measured at all numbered sites in Fig. 2.3 then it will be possible to identify the problem with magnification. On the other hand, measurements taken only at the sites numbered one will confound magnification and translation. The problem illustrated in Fig. 2.3 is a consequence of the non-random properties of overlay. Because of the physics of wafer steppers, overlay will vary in particular ways across exposure fields and between fields. There is a decided element of non-randomness associated with step-and-repeat and step-and-scan systems. Intrafield sources of variation — magnification in this example — repeat from field to field. Measurements taken in several fields across a wafer, but at a single intrafield location, will share an offset from the process mean that is characteristic of the intrafield error at those sites within the fields. Further discussion of the issue of measurement locations in the context of overlay will be deferred until Chapter 5, where the subject of overlay will be covered in detail.

Linewidths will also vary characteristically across exposure fields, as a consequence of lens aberrations and linewidth variations on the reticles. Measurements taken at only a few points within each exposure field, and involving only one or two feature types, is insufficient to characterize linewidths well. A particular feature on a reticle may have a linewidth in the middle of the overall linewidth distribution on one lens, but could, as a consequence of a different aberration signature, have relatively low or high dimensions when printed using a different lens. Although linewidth characterization is usually inadequate when few points are measured with exposure fields, it is impractical and unnecessary to measure many points per field on production wafers. Linewidth variations caused by lens aberrations are stable over time, so lenses can be characterized off-line, and the resulting data can be used in conjunction with in-line data collected during production. Electrical linewidth measurement methods are useful for collecting the large amounts of data required for detailed characterization of linewidth variations across exposure fields.[3] These techniques will be discussed further in Chapter 8.

Several resist processes, such as coating and develop, involve radially symmetric geometries. It can therefore be expected that linewidths should vary as a function of wafer radius.[4] Measurement sites should be distributed across wafers in a layout that will capture radial variation. An example is given in Fig. 2.4. However, the layout shown in Fig. 2.4(a) has a particular inadequacy. The area of an annulus on a wafer increases linearly with the radius. Sampling plans which are weighted uniformly to the wafer area they sample must have a linear increase in the number of sampling points as a function of the radius [Fig. 2.4(b)].

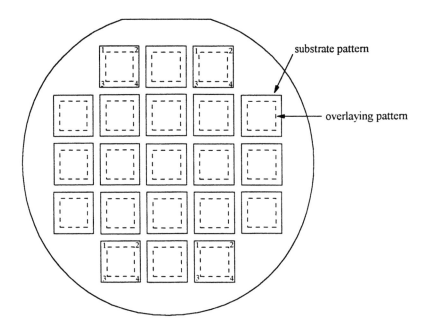

Figure 2.3. Overlay measurement sites.

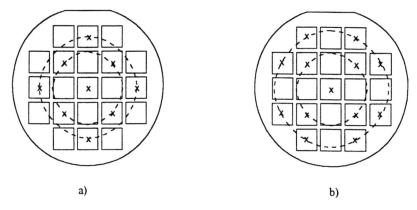

a)   b)

Figure 2.4. a) Measurement sites (marked X) on different radii. b) A sampling plan with improved weighting proportional to wafer area.

2.3 CORRELATIONS

At Sierra Semiconductor a number of years ago, the monitoring of resist thickness was assigned to third shift, the hours of midnight to 7 AM. Frequently, when the engineer responsible for the resist coaters arrived in the morning she would find the tracks disqualified for operation because the resist thickness was out of specifications. After taking care of important lots which were on hold, the engineer would process another resist thickness monitor to confirm the status of the coaters and would often find that the resist thickness was within specifications. This went on for several months, when the engineer was usually unable to reproduce the problem with resist thickness found by the night shift operator. At one point the engineer came into the fab at 5 AM to retrain the operator who monitored the resist thickness but found that she was doing everything correctly. Having eliminated human error as the source of the problem, attention was turned to possible physical mechanisms. Eventually, it was determined that the problem was related to the temperature control of the facility. The temperature in the fab was appreciably colder at night, when the operator on third shift monitored the resist thickness, compared to the temperature mid-morning when the engineer processed her monitor. This was an example of a correlation in time.

Correlations in time can be identified by calculating the sample autocorrelation function.[5] Suppose that a process is measured sequentially in time, resulting in values for a particular parameter: $x_1, x_2, x_3, \ldots$ Then the sample autocorrelation function is given by

$$r_k = \frac{\sum_{t=1}^{n-k} (x_t - \bar{x})(x_{t+k} - \bar{x})}{\sum_{t=1}^{n} (x_t - \bar{x})^2}, k = 0, 1, \ldots \qquad (2.22)$$

where $\bar{x}$ is the measured process average and n is the number of events which are considered. The problem observed at Sierra Semiconductor would have been observed with sampling intervals (the time between measuring x_t and x_{t+1}) equal to the duration of a single shift or shorter. The sample autocorrelation function is an estimate of the autocorrelation function ρ_k. Some properties of the autocorrelation function are:

$$|\rho_k| \leq 1 \text{ for all } k \qquad (2.23)$$

and

$$\rho_0 = 1. \qquad (2.24)$$

For a random function,

$$\rho_k = 0 \ \text{for} \ k \neq 0. \tag{2.25}$$

Significant deviations of ρ_k from zero indicate a non-random process. For a random process, the sample autocorrelation will vary with standard deviation $\dfrac{1}{\sqrt{n}}$, so values of $|r_k|$ significantly greater than $\dfrac{3}{\sqrt{n}}$ indicate that there are correlations in the process.

   The utility of Eq. 2.22 is illustrated by an example. A stepper engineer in one fab was particularly concerned about focus errors, and he instituted a focus check twice per shift for 50 consecutive shifts, generating a total of $n = 100$ measurements. The control chart which resulted is shown in Fig. 2.5. The engineer found that the "eight-in-a-row" Western Electric Rule was violated on several occasions. Further analysis indicated that this was the consequence of a situation in which successive measurements were correlated. In Table 2.2 the correlation coefficient (Eq. 2.22) is shown for the data of Fig. 2.5. Values of r_k for $k = 1,2,3$ are all greater than $\dfrac{3}{\sqrt{n}} = 0.3$, indicating that the data are correlated. This correlation leads to consecutive measurements that are close in value, thus leading to repeated violations of the "eight-in-a-row" Western Electric Rule.

k	r_k
1	0.657
2	0.461
3	0.302
4	0.141
5	0.153

Table 2.2. The sample autocorrelation function for the data shown in Fig. 2.5.

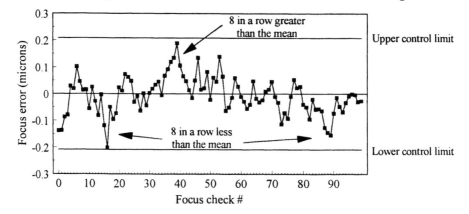

Figure 2.5. The control chart for frequently monitored stepper focus.

2.4 MEASUREMENT FREQUENCY

In order to lower the expense of control, the number of measurements should be minimized. At the same time, there must be a sufficient number of measurements to maintain control. The two requirements, expense of control and the costs that result from a loss of control, must be balanced. Accordingly, quick and easy measurements using inexpensive equipment can be done more frequently than measurements that take a long time and require costly tools. For example, defects on unpatterned resist films can be detected using laser scattering systems, in which a wafer can be measured in a few minutes. Tools which are capable of detecting defects on patterns are much more expensive than simple laser scattering systems, and it can take an hour or more to measure an entire wafer. Accordingly, resist defects can be monitored frequently, but economically, using laser scattering systems. On the other hand, there are many defect mechanisms that can only be detected on patterned wafers, so some level of inspection using the more expensive equipment will be required.

In general, measurements need not be made when processes can be expected to be unchanged. For example, the data in Table 1.2 show that resist thicknesses vary little during the time required to coat 10 consecutive wafers. Over longer periods of time considerably more variation is observed. Clearly, resist thickness is something that need not be monitored hourly, but significant variations may be observed with daily monitors.

The cost of losing process control depends on the type of circuits that are being manufactured. In a facility in which custom integrated circuits are being fabricated, a scrapped lot could represent 100% of the parts in the line for a particular customer. The cost of scrapping that lot is much more significant than the cost of a single scrapped lot in a DRAM fabricator. In general, one should not fear scrapping wafers, to the extent that preventive measures result in costs which are higher than the value of the scrapped wafers. Very often 100% inspection and measurement plans have been instituted to address problems that occur only once every few years. The expense of these inspections and measurements usually far exceeds the costs of the wafer scraps that were avoided. The objective should be to minimize costs.

2.5 SYSTEMATIC SOURCES OF VARIATION

Some parameters vary non-randomly. As noted previously, because of the geometry of resist processes, radial systematic variations across wafers are common. In such circumstances, statistics that characterize the process, such as the standard deviation, must be interpreted carefully. For example, consider the situation depicted in Fig. 2.6. A parameter varies linearly from 0 to A, and measurements are taken at equal intervals. The calculated sample standard deviation varies by nearly a factor of two as the number of measurements varies from three to more than 20. A similar situation occurs for parameters that vary quadratically (Fig. 2.7). Quadratic systematic variations are often observed when measurements are sampled along the diameter of a wafer that has parameters that vary radially.

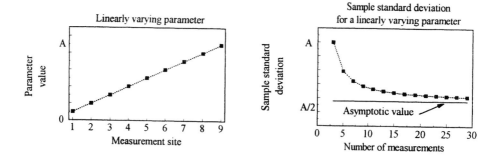

Figure 2.6. A linearly varying parameter, and the change in the sample standard deviation as a funtion of the number of measurements. The graph on the left illustrates a parameter sampled at nine evenly spaced sites.

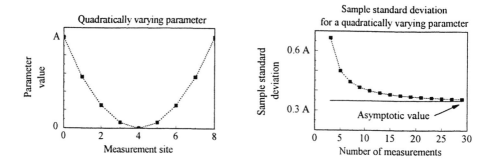

Figure 2.7. A quadratically varying parameter, and the change in the sample standard deviation as a function of the number of measurements. As illustrated on the left of the figure, the parameter is sampled at nine sites.

There can be significant consequences if systematic sources of variation are handled carelessly. For example, the number of measurements is frequently reduced as processes progress from development to the mature stages of manufacturing. As seen in Fig. 2.6 and Fig. 2.7, in the presence of systematic sources of variation it may appear that the variability of the process increases when the number of measurements is reduced. This would not be case if there was only random variation. In the absence of systematic sources of variation the average measured variation is independent of sample size. More than one lithography engineer has been asked to investigate apparent increased process variability that was only the consequence of a reduced sampling plan for a process with a significant systematic component.

As discussed in Section 2.2, reticles and lenses are sources of systematic variation. Linewidth variations and overlay signatures will repeat in every exposure field, but there will also be random components. The systematic errors complicate the statistical characterization of variation. One estimate of the distribution of errors, including the random and systematic contributions, can be obtained as follows. Suppose the random error $g(\varepsilon)$ at any point $\mathbf{z} = (x, y)$ on the wafer is approximately represented by a normal distribution about the mean error at $\mathbf{z}$:

$$g(\varepsilon, \mu(\mathbf{z})) = \frac{1}{\sigma\sqrt{2\pi}} e^{-\frac{(\varepsilon - \mu(\mathbf{z}))^2}{2\sigma^2}} . \qquad (2.26)$$

The distribution of errors overall is then given by

$$h(\varepsilon) = \frac{1}{A} \int_A g(\varepsilon, \mu(\mathbf{z})) \, dx \, dy , \qquad (2.27)$$

where the integrations are extended over the relevant areas of the wafers. In general, a non-normal distribution will result, but the fraction of sites $\phi(\varepsilon_L)$ that do not conform to specifications $[-\varepsilon_L, \varepsilon_L]$ can be computed using Eq. 2.28.[6]

$$\phi(\varepsilon_L) = \int_{-\infty}^{-\varepsilon_L} h(\varepsilon) \, d\varepsilon + \int_{\varepsilon_L}^{\infty} h(\varepsilon) \, d\varepsilon \qquad (2.28)$$

$$= 1 - \int_{-\varepsilon_L}^{\varepsilon_L} h(\varepsilon) \, d\varepsilon . \qquad (2.29)$$

As an example, suppose that the mean linewidth varies radially across a wafer:

$$\mu(r) = \frac{A}{r_0^2} r^2 , \qquad (2.30)$$

where r_0 is the radius of the wafer, and r is the radial distance from the center of the wafer. In addition to this systematic linewidth variation, at each point on the wafer the linewidth varies normally about the mean $\mu(r)$ with standard deviation σ. The resulting linewidth distributions are shown in Fig. 2.8 for various values for A (in units of σ) along a diameter. As A is increased the distribution is shifted to the right, since the mean is increased, and the distribution in broadened. The distributions also become skewed and flattened, but the values for skewness and kurtosis will remain within the confidence intervals shown in Tables 1.4 and 1.5, except for values of A much larger than shown in Fig. 2.8.

As discussed in this chapter, there are several issues that must be considered when determining the number and frequency of physical measurements, such as linewidths, resist thicknesses, and overlay. Having established some of the basic statistical issues with respect to process control, it is time to discuss the nature of process monitors specific to lithography in more detail. This discussion begins in the next chapter. The interplay between physical measurements and processing parameters will also be discussed. Further discussions of process drift and how to correct for it will be found in Chapter 7.

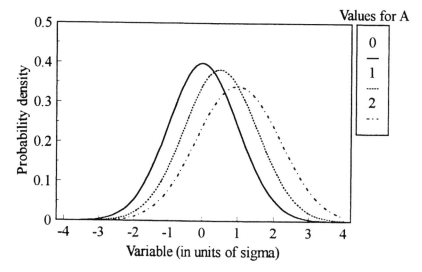

Figure 2.8. The probability density functions for three values of A in Eq. 2.31. There is a shift in the mean as a consequence of the quadratic variation in the mean across the wafer. The distribution also acquires a slight skew and becomes somewhat flatter for non-zero values of A.

[1] F. S. Hillier, " $\overline{X}$ - and R-Chart Control Limits Based on a Small Number of Subgroups," *Journal of Quality Technology*, Vol. 1(1), pp. 17 – 26 (1969).

[2] C. P. Quesenberry, "SPC Q Carts for Start-Up Processes and Short or Long Runs," *Journal of Quality Technology*, Vol. 23(3), pp. 213 – 224 (1991).

[3] C. Yu, T. Maung, C. Spanos, D. Boning, J. Cheung, H-Y Liu, K-J. Chang, and D. Bartelink, "Use of Short-Loop Electrical Measurements for Yield Improvement," *IEEE Trans. Semicond. Manuf.*, Vol. 8(2), pp. 150 – 159 (1995).

[4] R. C. Elliott, R. R. Hershey, and K. G. Kemp, "Cycle-time Reduction of CD Targeting using Automatic Metrology and Analysis," SPIE Vol. 2439, pp. 70 – 77 (1995).

[5] G. E. P. Box, G. M. Jenkins, and G. C. Reinsel, Time Series Analysis, 3rd Edition, Prentice Hall, Englewood Cliffs, NJ (1994).

[6] T. R. Groves, "Statistics of Pattern Placement Errors in Lithography," J. Vac. Sci Technol., B9(6), pp. 3555 – 3561 (1991).

CHAPTER 3
SIMPLE AND COMPLEX PROCESSES

3.1 DEFINITIONS

The purpose of statistical process control is the identification of abnormal variations in the materials, equipment, parameters or procedures used as inputs for particular processes (see Table 3.1). This is usually accomplished by measuring physical objects on the output, such as linewidths or overlay structures. When processes require test wafers, which often occurs during process development or the initiation of a new manufacturing process, the output of the process is decoupled from the input, and statistical process control cannot fulfill its primary purpose.

Input parameter	Examples
Materials	Photoresist, developer, anti-reflection coating, substrate films
Equipment	Particular stepper model, resist track
Procedure	Operating specifications, linewidth target and allowable range
Parameters	Exposure dose, overlay offset, focus setting

Table 3.1 Process input parameters and typical examples.

This situation is shown schematically in Fig. 3.1. The inputs collectively comprise the process. A single wafer is taken from a lot of wafers and processed through the lithography operation. After the processing is complete, this test wafer is measured for parameters of interest, such as linewidths or overlay. From the values of these measurements, in comparison to the process targets, the remaining wafers in the lot are processed through the lithography operation with adjusted process parameters. For example, the exposure dose might be adjusted to bring linewidths to the process target. Changes in the dose might be required to compensate for drift in the stepper's dose control system or the changes in the resist process. Measurements of the linewidths of the lot, except for the test wafer, will not reveal a drift in the stepper's dose control system or a change in the resist process, because the exposure dose has been adjusted to compensate for these instabilities. In order for statistical process control to reveal variations and instabilities in the inputs, it must be applied to a simple process, where the input variables are directly coupled to the measurable output.[1] In this chapter a process control methodology applicable to situations in which test wafers are used is presented.

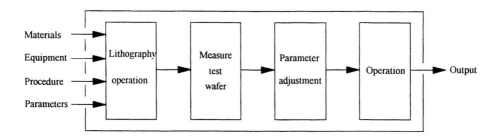

Figure 3.1. Material flow in a complex process involving test wafers.

3.2 WHY TEST WAFERS ARE USEFUL

The use of test wafers is common in lithography, particularly for processes in development or the early stages of manufacturing. Exposure doses are changed to adjust critical dimensions, stepper alignment offsets are adjusted to improve overlay, or a spin speed might be changed to adjust resist thickness. This practice of making adjustments based upon a small set of data, while common, is contrary to the principles of statistical process control. The idea that adjustments to processes are in conflict with statistically sound practices was preached by the great W. Edwards Deming, who illustrated the problem through his funnel experiment.[2]

The funnel experiment is conducted by dropping a marble through a funnel, which is held above a table which has a target marked at a single point (Fig. 3.2). A marble is dropped through the funnel. After the marble lands on the table it will bounce, roll and eventually come to rest. The distance between the target and the final resting place of the marble is measured. This process is repeated, with the objective of producing the tightest cluster about the target of finally resting spots. Four rules for the process are proposed:

Rule 1. Leave the funnel fixed, aimed at the target, and make no adjustments.

Rule 2. At drop k, the marble will come to rest at point $\vec{x}_k$. Move the funnel to $-\vec{x}_k$ from its last position for drop $k+1$.

Rule 3. Set the funnel over point $-\vec{x}_k$, measured from the target, for drop $k+1$.

Rule 4. For drop $k+1$, set the funnel over the previous spot where the marble came to rest after drop k.

Rules 2 – 4 are procedures for adjusting the process in attempts to tighten the distribution about the target. The results of operating according to these rules are:

Rule 1. This rule produces the tightest cluster of resting spots about the target.

Rule 2. The output from this process is stable, but the variance is twice that of Rule 1.

Rule 3. This system is unstable. The marble will eventually move further and further from the target.

Rule 4. This system is also unstable.

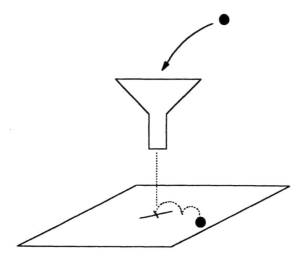

Figure 3.2. The funnel experiment.

The funnel experiment was designed to illustrate the futility of making adjustments based on the outcome of a randomly varying process, since all attempts to improve the process had the opposite effect. On the other hand, experience with test wafers has shown that improvement is possible using data from only one wafer. This apparent contradiction can be understood from the nature of the sources of variation that occur in lithography. Adjustments typically are applied soon after test wafers are processed and measured. Consequently, long-term drift is not a factor which affects the applicability of test wafer data to the rest of the wafers in the lot.[3] There are several physical phenomena that lead to long term drift. For example, the overlay of wafer steppers is affected by the temperature of the stepper. In spite of efforts to control the equipment, the stepper's temperature will vary by small amounts. However, because of the large thermal mass involved, this temperature change can occur only very slowly. Effects which result from changes in barometric pressure will also vary at the rate the pressure changes significantly, which is typically over the course of several hours.

Linewidths and overlay are also affected by the substrates, which are often more repeatable within a lot of wafers, compared to the lot-to-lot variation. Test wafers capture the substrate signature of the lot. There are also events that are identifiable, which indicate that the process has changed. This includes new batches of photochemicals, equipment maintenance, and new reticles.

The funnel experiment was designed to demonstrate the futility of making adjustments to processes in a crudely reactive mode. Sources of variation that are not identified or subjected to control will apparently lead to "natural" levels of variation. There is considerable difference between identifying and controlling sources of variation and adjusting processes based solely upon prior outcomes. There are no sources of variation, at the level of significance for semiconductor processing, that are inevitable and cannot be eventually subjected to control. The decision to establish control is based upon knowledge of the sources of variation and an assessment of cost-effectiveness.

3.3 HOW TO ADDRESS COMPLEX PROCESSES IN LITHOGRAPHY

In lithography, the use of test wafers is usually justifiable, and often helpful. However, when they are used, statistical process control cannot be applied conventionally for controlling the process. Without one of our most powerful quality improvement tools, it becomes difficult to improve the process so that test wafers can be eliminated and efficiency improved. In this section, a method will be presented for applying statistical process control to complex processes. This method is a tool for process monitoring and improvement.

Conventionally, processes are monitored and controlled by trending and analyzing physically measurable data such as linewidths or overlay. We have seen the difficulties in applying this method to complex processes. Instead of monitoring physically measurable data, complex processes may be controlled by trending and analyzing process parameters[4] (Fig. 3.3). In this approach, the process parameters used to process each lot are trended. For example, one of these parameters could be the exposure dose. For a completely stable process the exposure dose would be constant. Variations in resist materials, exposure tools and substrates would require changes in the exposure dose in order to achieve linewidths that are close to the target value. Fluctuations in the exposure doses used to process lots would reflect these variations.

The basis for this approach can be appreciated analytically. Consider a situation involving critical dimensions. Linewidths are determined by a number of factors, such as exposure dose (E), resist thickness (t_{resist}), post-exposure bake temperature (T_{PEB}), and so forth. Conceptually, the linewidth is a function of these factors:

$$\text{linewidth} = x = f\left(E, t_{resist}, T_{PEB}, \ldots\right) \tag{3.1}$$

even though the exact form of the function f may not be known. It should be noted that to the level of precision relevant to microelectronics processing, this equation is deterministic, a consequence of the natural laws of physics and chemistry.

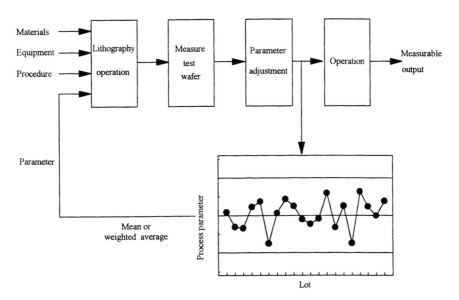

Figure 3.3. Process parameter monitoring. Shown in this example is a feedback loop.

If all of the independent variables (E, t_{resist}, T_{PEB}, ...) could be held exactly at their set points, the linewidth x would equal its target value x_o. Variations in x are a consequence of fluctuations in the values of E, t_{resist}, T_{PEB}, and the other independent parameters about their set points. For example, a hotplate may be set to a particular temperature, but the actual temperature will vary about its set point, and linewidths will vary as a consequence of this variation in temperature. Monitors of the linewidth, where the equipment settings for all process parameters are set to fixed nominal values, will show variations that indicate that one (or more) of these independent variables is fluctuating. Monitors of the linewidth cannot reveal uncontrolled fluctuations of independent variables if one of these parameters is varied intentionally, such as when the exposure dose is changed on the basis of test wafer measurements, unless one can predict exactly how the change in dose will vary the linewidth and compensate accordingly. On the other hand, Eq. 3.1 can be inverted, at least conceptually, to express the exposure dose in terms of a function f' of the other independent parameters and the linewidth.

$$E = f'\left(x, t_{resist}, T_{PEB}, ...\right) \tag{3.2}$$

When the linewidth x in Eq. 3.2 is held at its target value x_o, the right side of this equation consists entirely of factors which are nominally constant, i.e., none of these parameters are varied intentionally. This is the situation that arises when test wafers are used. The linewidth *target* x_o is held constant, as are the settings for all of the other parameters, and it is the exposure dose on the left-hand side of the equation that is varied. Consequently, variations in the left side of the equation indicate real fluctuations in the factors on the right-hand side about their set point values. Monitors of E can thus be used to monitor and control the process.

Alternatively, the control methodology can be based on physical measurements from the test wafers rather than the processed lots, so long as the test wafers are always processed using the same equipment set point values. In this case the measurements do reflect real variations of the inputs. However, there is an advantage to the approach where process parameters, such as exposure doses, are trended. Because it is desirable to minimize rework, the monitoring of process parameters provides the feedback for adjusting the process inputs to maximize the likelihood that the test wafer will be within specifications[5] (shown in Fig. 3.3). In fact, it was in the context of reducing rework and test wafers that this methodology was first used.[6]

This approach to process control, by monitoring operating parameters such as exposure doses, addresses the fundamental problem introduced by test wafers, the decoupling of the quantity that is tracked from the process inputs. However, there are a number of problems with this approach that need to be addressed. First, adjustments to operating parameters are not always made. When the results of test wafers are very close to target, there may be a reluctance to adjust the process parameters. As a consequence, the distribution of process parameters used on product lots will look like the histogram shown in Fig. 3.4. Many lots are processed at a single value, while there are no lots that are processed with the process parameter near this value. The gaps result from lots which have test wafer results near the target. The distribution shown in Fig. 3.4 is very non-normal, and the data need to be analyzed accordingly. In particular, the Western Electric Rules cannot be applied to these data directly.

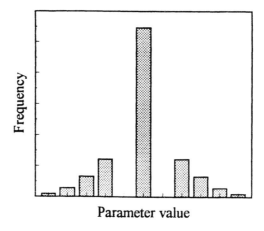

Figure 3.4. Histogram of a process parameter used to process lots. The ordinate is the process parameter used to process the lots.

There are also some problems in targeting the parameters used for processing lots. When single test wafers are used there is inevitable statistical noise which results from wafer-to-wafer variation. Moreover, even though the short-term drift is significantly less than the long term drift, there still may be sufficient drift to affect the targeting of operating parameters. Finally, the relationship between the physical parameters and the operating parameters must be taken into account properly. Consider, for example, critical parameters as a function of exposure doses (Fig. 3.5). Over a sufficiently large range, this function is non-linear. In manufacturing, there is a trade-off between simplicity and accuracy, and a linear approximation is often used. The relationship between linewidth and exposure dose is affected by other parameters, such as focus. Consider the curves in Fig. 3.6. A smaller change in dose is required to change the spacewidth by a certain amount at non-zero defocus, compared to the dose change required at zero defocus. If a test wafer is processed with some amount of defocus, the remainder of the lot will not be targeted correctly if the amount of change in spacewidth is estimated on the basis of data collected at zero defocus.

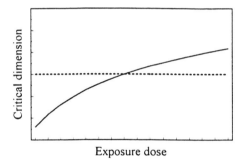

Figure 3.5. Linewidth as a function of exposure dose, showing typical non-linearity.

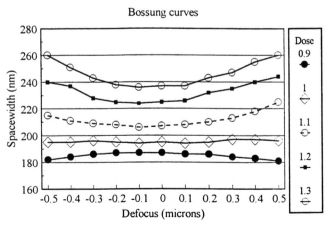

Figure 3.6. Spacewidth variations as functions of dose and defocus. The spacewidth varies more with dose at larger values of defocus.

Because of the various problems with the method of monitoring process parameters, a more sophisticated form of the method is used where there are good data handling capabilities. In this approach, the lot data are also used to refine the estimate of what the process parameter should have been[7] (Fig. 3.7). Highly automated data handling is required, since the process parameters that are trended are not simply the parameters used for the lot, but they are corrected using measured data after the lot is processed. This increases the amount of data on which the trended parameter is based, since multiple wafers are often measured after the lot has completed processing. This method involves significant quantities of data and must be automated, particularly since process parameters can be specific to each combination of product, masking layer, and product. Small corrections will be included automatically, correcting the deficiency illustrated in Fig. 3.4 without making the job more complicated for operators.

Automatic feedback should always be applied with caution, since adjustments are being made in one parameter to compensate for variations in another. This is generally acceptable, within limits, for certain types of corrections, such as using exposure dose to compensate for variations in the resist processing. On the other hand, it would be best to correct a change in focus rather than compensate for its effect on linewidth by changing the exposure dose. Active process control will be revisited in Chapter 7.

3.4 Distinguishing between Layer-Specific and Equipment-Specific Effects

Because of the immaturity of film deposition operations, anneals, etch processes, and polishing, substrates may be quite variable when bringing up new manufacturing facilities. This will lead to large fluctuations in linewidth and overlay. At the same time, there will be little confidence in the stability of the lithography process itself, because the lithography operation will be similarly immature. This problem is also quite common on development pilot lines, where non-lithography processes, which can have profound effects on lithography, are changed frequently. Even in mature manufacturing operations there can be losses of process control that can cause substrate variability. Lithographers need a method to distinguish between variations in substrates and fluctuations caused by variable lithography processes and equipment.

A method capable of accomplishing this involves monitors of the lithography equipment and processes that employ extremely reproducible substrates. For most applications, bare silicon wafers work fine. For example, a silicon wafer can be processed through the lithography operation daily, or some other consistent interval, and the linewidths can be measured. These measurements provide a baseline indicating the fundamental stability of the lithography operation. Variations seen on such monitors can be expected on product wafers. If these monitors are relatively stable, then large variations on product wafers can be attributed to some mechanism involving the substrate.

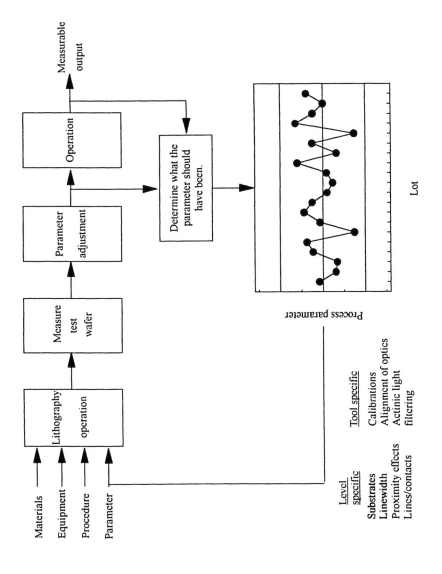

Figure 3.7. Feedback loop showing the parameter adjustments based on measurements from the lots.

An example of such a monitor is shown in Fig. 3.8, which was used to monitor linewidth control on early deep ultraviolet (DUV) excimer laser steppers.[24] These steppers were being used on a development pilot line. At the same time, the photoresists were undergoing development, with large variations in photospeed between batches. The steppers were very immature, and there was a steep learning curve that needed to be climbed. Because the substrate composition was changing frequently as a result of the development function of the pilot line, large linewidth variations could be expected. These needed to be distinguished from variations in resist materials and lithography equipment. The bare silicon image process monitor enabled the engineers to monitor the process, identify changes, and correct them prior to processing "product" wafers.

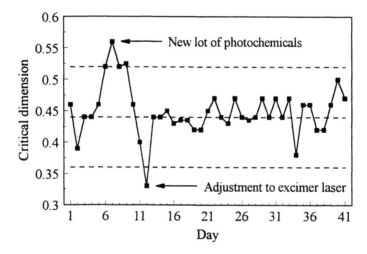

Figure 3.8. Linewidth monitor for an early excimer laser photocluster.

Patterned wafers that are measured on linewidth measurement tools provide good monitors for the entire imaging process. The measurements often require expensive equipment and can take a long time. Photoresist processing equipment and steppers can be left idle while measurements are being taken. Measurements of E_0, the minimum exposure dose to clear positive resists to the substrates, provide quicker monitors that do not require expensive and time-consuming metrology, and which provide most of the information needed to control resist processes.[8] For these types of monitors, "open frame" exposures are used. These are large area exposures, typically several millimeters per side, often accomplished by having no reticle in the stepper. The dose at which the resist is first cleared from the substrate can be determined from visual inspection. Measurement noise can be reduced and interpolation between doses is possible by measuring resist thickness at each exposure and fitting the data to a common resist contrast curve, with only a small increase in time. Nearly all positive photoresists have similar resist contrast curves, where resist thickness varies linearly as the logarithm of the dose, in the vicinity of E_0:

$$\text{thickness} = T_0 \gamma \ln\left(\frac{E_0}{E}\right), \tag{3.3}$$

where T_0 is the initial resist thickness, γ is the resist contrast, and E is the exposure dose. Measurements of E_0 have been used to identify fluctuations in materials and processes.[9] Linewidth measurements provide better monitors for very high contrast resists, such as many DUV materials, because E_0 is very hard to measure for these resists.

Similarly, "golden" or "holy" wafers can be used for controlling overlay.[10,11] Like the wafers used for monitoring linewidths, these should be well controlled substrates. Generally, they consist of patterns etched into substrates. The substrates should be chosen so that the alignment targets and overlay measurement structures are free from noise factors, such as grains in films, that will degrade the utility of these wafers for monitoring the status of the equipment. Quite frequently, patterns etched into bare silicon are ideal. Bare silicon substrates also do not expand or contract from being etched, as might occur from etching deposited silicon nitride or oxide films, and this enhances the reproducibility of the substrates. Wafers will expand or contract because of changes in temperature, regardless of composition, and any overlay control scheme must take this into account. Silicon has a coefficient of thermal expansion of 3 ppm/$^{\circ}$C.[12] Points on opposite sides of a 200 mm wafer will move 60 nm apart for every 0.1°C change of temperature. This represents a substantial fraction of overlay budgets for 250 nm technologies and smaller.

Equipment and processes may drift, and these changes may not be observable on monitors using ideal substrates. For example, an alignment system may drift, causing difficulty aligning wafers whose alignment targets provide low signal-to-noise. This drift may not be observable for targets that provide strong alignment signals, as should exist on the "golden" wafers. Similarly, problems in resist processing may occur only for certain substrates, or wafers with topography, and therefore may not be observed on linewidth monitors that use bare silicon substrates. Although there may be process problems that may go undetected using process monitors that employ particular ideal standardized substrates, these types of monitors remain extraordinarily useful. When there are problems with overlay, there is no substitute for the knowledge that the steppers are (or are not) providing good overlay on standards which, if not etched into stone, are at least etched into silicon.

Processes change when operating parameters are adjusted, and these changes must be taken into account when monitoring processes. This is particularly important in lithography, where test wafers are often used, and process parameters are adjusted on the outcomes of test wafer measurements. A method for addressing alterations in process parameters was presented in this chapter. It is also important to have a capability for distinguishing between changes in the non-lithographic parts of the semiconductor manufacturing processes and those that occur in the lithography process itself. The use of "golden" wafers provides this capability. One can take greater advantage of these techniques by going beyond purely mathematical analyses of the data and using knowledge of the physics and chemistry specific to lithography to more fully interpret the results. We now turn to the science of lithography.

[1] H. J. Levinson, "Control and Improvement of Complex Processes," *Quality Engineering*, Vol. 5(1), pp. 93 – 106 (1992).

[2] W. Edwards Deming, *Out of the Crisis*, MIT Center for Advanced Engineering Study, Cambridge, MA (1982).

[3] J. F. MacGregor, "A Different View of the Funnel Experiment," *Journal of Quality Technology*, Vol. 22, pp. 255 – 259 (1990).

[4] L. C. Mantalas and H. J. Levinson, "Semiconductor Process Control," in *Handbook of Critical Dimension Metrology and Process Control*, K. M. Monahan, ed., SPIE Press, Bellingham, WA (1994).

[5] M. Drew and K. Kemp, "Automatic Feedback Control to Optimize Stepper Overlay," SPIE Vol. 1926, pp. 422 – 428 (1993).

[6] H. J. Levinson and C. DeHont, "Leading to Quality," *Quality Progress*, pp. 55 – 60 (1992).

[7] C. P. Ausschnitt, A. C. Thomas, and T. J. Wiltshire, "Advanced DUV Photolithography in a Pilot Line Environment," *IBM J. Res. Develop.*, Vol. 41(1/2), pp. 21 – 36 (1997).

[8] D. Heberling, "Litho Equipment Matching with E_0," Proceedings of the KTI Microelectronics Seminar, pp. 233 – 243 (1990).

[9] C. Takemoto, D. Ziger, W. Connor, and R. Distasio, "Resist Tracking: A Lithographic Diagnostic Tool," SPIE Vol. 1464, pp. 206 – 214 (1991).

[10] M. A. van den Brink, C. G. M. de Mol, H. F. D. Linders, and S. Wittekoek, "Matching Management of Multiple Wafer Steppers Using a Stable Standard and a Matching Simulator," SPIE Vol. 1087, pp. 218 – 232 (1989).

[11] K. Kemp, C. King, W. Wu, and C. Stager, "A "Golden Standard" Wafer Design for Optical Stepper Characterization," SPIE Vol. 1464, pp. 260 – 277 (1991).

[12] *Handbook of Chemistry and Physics, 61st Edition,* CRC Press, Boca Raton, Florida (1981).

CHAPTER 4
LINEWIDTH CONTROL

In the preceding chapters the focus was primarily on some of the mathematical tools available for controlling processes. Statistical process control is a methodology that enables people to identify abnormal levels of variation. Once it has been determined that a process is out of control, the sources of the excessive variation or process drift need to be identified and corrected. Moreover, it is preferred to have processes that normally have low levels of variation. In order to design processes which are intrinsically stable or to take corrective action when processes go out of control, it is necessary to understand and apply the science of lithography. For example, when linewidths drift, one needs to understand what parameters can potentially cause changes in the process.

Lithographers need to control linewidths, overlay and defects. Aspects of lithographic science relevant to linewidth control will be discussed in this chapter. Similar discussions for overlay will follow in Chapter 5, and for yield in Chapter 6.

4.1 CAUSE AND EFFECT

There are a large number of variables that can affect linewidths, many of which will be discussed in the next section of this chapter. If control of a process is lost, it is the job of the process engineer to determine what specifically caused the abnormal change, among all possible causes, so that appropriate corrective action can be taken. A technique for identifying probable causes, particularly useful for brainstorming with groups, is the cause-and-effect diagram.[1] The underlying principle of this technique is causality, i.e., there is a cause for every observed phenomenon.

An example of a cause-and-effect diagram is shown in Fig. 4.1. The cause-and-effect diagram is often called the Ishikawa diagram, named for its inventor, and is sometimes referred to as a fishbone diagram, because of its appearance. In the cause-and-effect situation, the effect is the loss of process control. There can be a number of possible causes for this. For manufacturing, a universally applicable cause-and-effect diagram has been constructed, shown in Fig. 4.2. All losses of control can be traced to one of the six categories represented in Fig. 4.2.

Having identified certain general classes of causes, it remains to determine the cause of poor process control more specifically. To this end, more detailed causes can be linked to the general category, in the graphical format shown in Fig. 4.3. Suppose post-exposure bake (PEB) temperature is suspected as being the cause of a change in linewidths. There are a number of reasons that the PEB

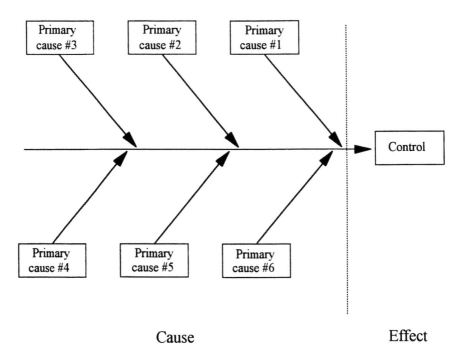

Cause   Effect

Figure 4.1. The basic form of the cause-and-effect diagram.

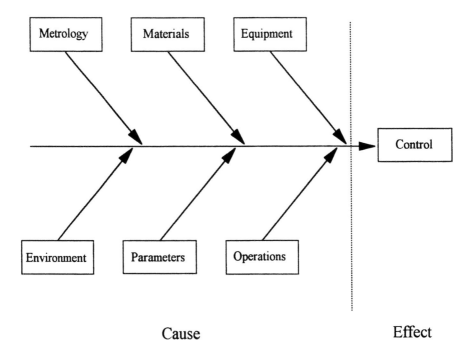

Cause   Effect

Figure 4.2. A universal cause-and-effect diagram for manufacturing.

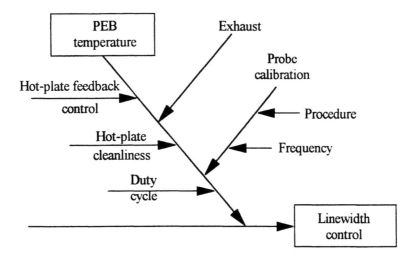

Figure 4.3. A branch of the cause-and-effect diagram, for the problem of linewidth control.

temperature may have changed, and some possibilities are shown. Some of the suspected causes may have more specific underlying origins, and these can be shown as well.

The cause-and-effect diagram is qualitative, in that it does not distinguish relative magnitudes of effects due to specific causes. After collecting possible causes, engineers should decide which causes are most likely and most significant, and take appropriate measurements to determine if these are indeed the primary reasons that control has been lost.

Process control is often lost in high stress situations. Under such circumstances there is a tendency for fingers to be pointed at people rather than technical factors. The cause-and-effect diagram is useful for refocusing engineers and managers on the technical problems instead of people. The human aspect of process control will be considered further in Chapter 9.

4.2 INDEPENDENT VARIABLES

As expressed in Eq. 3.1, linwidths are functions of particular process variables. The most significant of these process factors are listed the Table 4.1. In addition to these variables, there are other factors that are more specific to particular processes. Consider, for example, a puddle develop process. A puddle process is one in which developer is dispensed over the photoresist, thereby forming a puddle. After a preset period of time the developer is removed from the wafer, usually by using the centrifugal force of spinning. The resulting linewidths will depend upon the volume of developer dispensed, whether it is a single or multiple puddle process, the speed at which developer is applied, and the dispense nozzle configuration. The remainder of this chapter will be devoted to a more detailed discussion of the parameters in Table 4.1, and the ways in which they are known to affect linewidths.

Exposure dose	Developer normality
Focus	Developer time and temperature
Resist thickness	Substrate reflectivity
Softbake time and temperature	Resist composition
Post-exposure bake time and temperature	Ambient amine levels (DUV resists)

Table 4.1. Important variables which affect linewidths.

It is essential to know quantitatively how each process parameter affects linewidths and what the normal levels of variation are for each parameter. For example, suppose that the post-exposure bake changes linewidths by 2 nm/°C. With this knowledge it is possible to conclude that an 0.5°C change in the temperature of the hotplate used for the post-exposure bake cannot account for an observed 10 nm shift in the mean linewidth, i.e., even though the hotplate temperature did change, the magnitude of the change was insufficient to account for an observed linewidth shift. It is also essential to know facts such as the normal 3σ variation of hot-plate temperatures. In general, the relationship between the change in a critical process parameter, such as linewidth, and changes in parameters, such as bake temperatures, can be expressed using Taylor's theorem. Let the function f represent the critical process parameter, and x the independent variable. If x_o is the nominal operating point for x, then:

$$f(x) = f(x_o) + \frac{df}{dx}(x - x_o) + \frac{1}{2}\frac{d^2f}{dx^2}(x - x_o)^2 + \dots \,, \qquad (4.1)$$

where the derivatives are evaluated at $x = x_o$. The change in the critical parameter, as a consequence of a change in the independent variable from its nominal value, is:

$$f(x) - f(x_o) = \frac{df}{dx}(x - x_o) + \frac{1}{2}\frac{d^2f}{dx^2}(x - x_o)^2 + \dots \qquad (4.2)$$

In many circumstances only the first term on the right-hand side needs to be retained, and there is a linear relationship between the linewidth and the independent variable. To understand the change in the linewidth, one needs to know the magnitude of $\frac{df}{dx}$, which is the rate of change (e.g., 2 nm/°C), while $(x - x_o)$ is the amount the independent variable changes (e.g., 0.5°C). If x varies normally about x_o with standard deviation σ, then f will also vary normally, with standard deviation $\dfrac{df}{dx} \cdot \sigma$, when f and x are related linearly. In a number of

circumstances, $\dfrac{df}{dx} \approx 0$. This occurs, for example, when x represents focus.[2] In this situation the relationship between the linewidth and the independent variable is more complicated, but tractable, since the methodology just described can be applied using the next higher order terms in the expansions Eq. 4.1 or 4.2. This more complicated relationship arose in the earlier discussion on skewed distributions (Section 1.4). A software package, NORMAN, has been developed at IMEC[3] that combines the deterministic relationship between linewidth and process parameters — such as focus, dose, resist thickness, and reticle dimensions — and the probabilistic variation in these very parameters, to produce histograms of linewidth.[4, 5]

Such modeling software is very useful for process improvement, because it can be used to rank the sources of variation quantitatively.[6] Plots of the sources of variation in the format shown in Fig. 4.4 are referred to as Pareto Charts.[1] In the Pareto Chart, contributors are ordered from left to right according to the magnitude of their contribution. In order to do this, contributions must be characterized according to a common metric. This type of analysis is very useful, because it shows clearly the improvements that will reduce linewidth variation most significantly. Those who manage lithography operations or engineering projects should use Pareto Charts in order to focus resources in those areas where improvement will have the biggest impact.

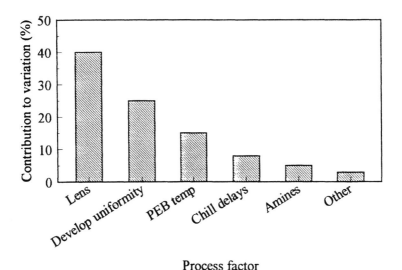

Figure 4.4. An example of a Pareto chart.

There is one type of process change that is highly desirable — improvement. Better processes result from identification of the causes of variation, and the application of controls to the sources. Because resources are limited, improvement efforts should be focused on the largest sources of variation. Controlling these will provide the greatest improvement to the overall process.

Models can also be used for improving quality by indicating that the accounting of sources of variation is incomplete. Independent measurements of sensitivities to parameters such as resist thickness and focus, and the amount of variation normally seen for these parameters, can be used as inputs to a model for calculating total linewidth variation.[7] If the model produces less variation than the actual process, then some significant source of variation has been overlooked. Processes may be improved by identifying and controlling this factor.

Additional engineers and technicians and new equipment are often required in order to improve process control. Since these cost money, it is essential that managers be prepared to justify increased expenditures. The cost of inadequate process control, due to rework and lost yield, can be compared to the costs of hiring additional staff or purchasing new equipment, if the quality costs have been estimated in monetary terms.[8]

If sufficiently large variations in parameters occur, the utility of Taylor expansions, such as Eqs. 4.1 and 4.2, is reduced. Consider the variations in linewidth shown in Fig. 4.5, which were calculated using the lithography simulation program, PROLITH2.[9] If the resist thickness varies by an amount in excess of the quarter-wavelength of light, the functional form of linewidth variation goes considerably beyond the first few terms in a Taylor series expansion. If a particular process represents such a situation, then the concepts in this section can be extended by accounting for the more complicated functional dependence of linewidth on particular parameters. Processes which are centered at a minimum or maximum of the swing curve, and where the resist thickness varies less than a quarter-wavelength of light, will have skewed distributions as a consequence of thin film thickness variation.

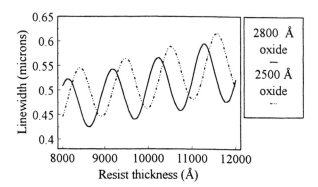

Figure 4.5. Variations in linewidths as a function of resist thickness, calculated using PROLITH 2. The nominal linewidth was 0.5 microns, and the optics were *i*-line. The substrate was oxide on silicon.

While each of the process factors summarized in Table 4.1 affect linewidths in quantifiable ways, each parameter is itself the consequence of additional parameters, most of which are intimately related to equipment and materials. For example, bake time and temperatures will be functions of hotplate temperature control, its response time and calibration, and drift. Ambient temperatures and exhaust, as well as wafer backside contamination also affect the effective bake temperature. Process control requires an understanding of the specific equipment and materials used.

4.2.1 Exposure dose

The amount of light incident on resist-coated wafers is controlled on steppers, using closed-loop feedback control (Fig. 4.6). Dose repeatability on g- and i-line wafer steppers is typically $< \pm 1\%$, and $< \pm 1.5\%$ on DUV systems. The absolute control of dose is limited by calibration errors, which are currently about 2% for 248 nm excimer laser illumination.[10] Generally, linewidths remain within specifications as long as the amount of light absorbed in the resist varies by no more than $\pm 10\%$, so the level of dose control provided by wafer steppers is quite adequate for 250 nm and older technologies. Improved levels of dose control will be required for optical lithography well below the wavelength of light.

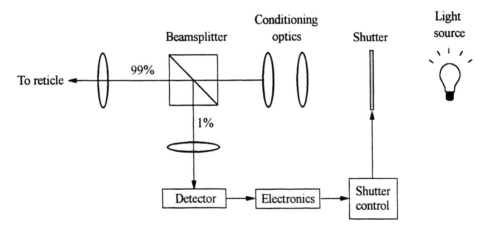

Figure 4.6. The amount of light exiting the illuminator is monitored constantly, and a closed-loop system ensures that the stepper produces a controlled exposure dose independent of lamp or laser brightness.

The amount of light absorbed by resist is determined by the light reflected by the substrate as well as the incident light. Consider a situation involving silicon wafers with uniform oxide films on them. If the wafers are coated with a photoresist film of varying thicknesses and then illuminated with uniform exposures of light, the total amount of light absorbed by the resist films is shown in Fig. 4.7. Because of thin film optical interference effects, the amount of light absorbed varies by ± 24% as the resist thickness varies over the range of thicknesses shown in the graph. Because of the manner in which the dependent variable "swings" up and down as a function of the film thickness, graphs such as those in Fig. 4.7 are referred to as swing curves.

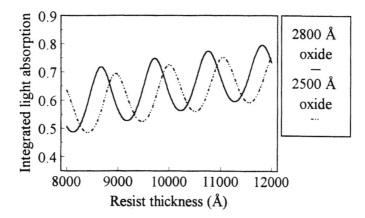

Figure 4.7. The amount of light absorbed by a resist film under uniform *i*-line illumination. This is the same configuration for the curves shown in Fig. 4.5.

It should be noted that swing curves obtained with particular numerical aperture and illumination settings may not apply in situations where different values for these parameters are used.[11] This may be understood as follows. The maxima and minima in Fig. 4.7 occur when particular phase relationships are satisfied between the incident and reflected light (Fig. 4.8). These relationships will depend on the optical path lengths of the light in the photoresist film, and these will be different for non-normal light at different angles of incidence (Fig. 4.8b).

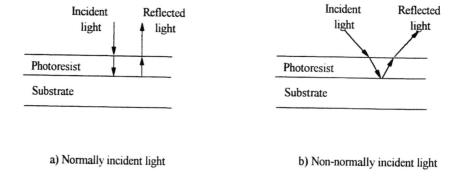

a) Normally incident light   b) Non-normally incident light

Figure 4.8. Incident and reflected light in photoresist films. Light has different path lengths for normally and non-normally incident light. Consequently, the incident and reflected light have different phase relationships depending upon the angle of incidence.

High numerical aperture, off-axis illumination, and the pattern pitch will all affect the angles of incidence and hence the resist thicknesses at which swing curve extrema will occur. This can be understood by considering the imaging of a simple diffraction grating. Consider a beam of light that passes through a grating consisting of long, parallel lines and spaces. If the light comprising this beam is all of one wavelength, and all of the waves in the beam are in phase, then the light will diffract into well-defined beams[12] (Fig. 4.9). Suppose the incident beam is in a plane perpendicular to the lines of the grating. Then the diffracted beams will be transmitted at angles θ given by

$$\sin\theta - \sin\theta_0 = \frac{m\lambda}{p} \tag{4.3}$$

where p is the pitch of the grating, θ_0 is the angle of incidence, m is an integer, and λ is the wavelength of the light. This physics has implications for imaging in lithography. Depending on the angle of incidence of the illumination and the pitch of the pattern, the light from a grating pattern projected through a lens will have different angles of incidence θ on the wafer. The path length through the photoresist will depend on θ, as will the exact positions of the minima and maxima on the swing curves. Consequently, the positions of swing curve extrema will be functions of the pitch, illumination conditions, and the numerical aperture.

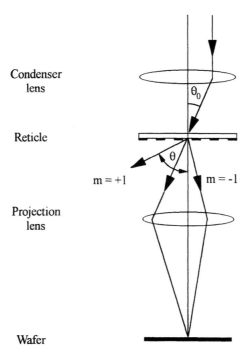

Figure 4.9. A beam of light passing through a grating pattern.

Large linewidth variations can also occur when the thickness of the oxide underlying the photoresist is changed by only a few hundred angstroms. From the graph shown in Fig. 4.5 it is evident that a process will have, for a particular oxide thickness, the least linewidth variation as a consequence of changes in resist thickness if the target value for the thickness of the resist is either a maximum or minimum of the appropriate curve. If the oxide thickness is well controlled, then this is sufficient. There are circumstances, often the result of chemical-mechanical polish, where the oxide thicknesses can vary considerably, and there is no stable operating point for the resist thickness. In such a circumstance, something is needed to address the fundamental cause of the swings in the graphs of Figs. 4.5 and 4.7, which is the reflectivity of the underlying silicon substrate.

Bottom anti-reflection coatings reduce the reflectivity of the effective substrate, while top anti-reflection coatings reduce the phase-coupling of incident and reflected light,[13] thus reducing the magnitude of the swings shown in Fig. 4.5. Anti-reflection coatings do increase process complexity and cost, so there has been some reluctance in using them. However, it is often true that the resulting improvement in linewidth control provides greater value than this cost, and the optimal solution is often now found to involve the use of anti-reflection coatings.

4.2.2 Resist thickness

In order to maintain operation at a swing curve extremum, resist thickness must be well controlled. A number of parameters modulate resist thickness and uniformity, and these are listed in Table. 4.2. The specific values for these parameters will depend upon the specific equipment and resists that are used. Nevertheless, it is very useful to know which parameters can be relevant to resist coating uniformity and control.

Temperatures play a significant role in resist coating process. Through the use of a chill plate prior to resist coating, wafers are brought to a controlled temperature. Resist thickness non-uniformity characterized by thick resist in the middle of the wafer can often be improved by raising the temperature of this chill plate. The ambient air will be at a certain temperature, which is usually controlled and independent of the chill plate temperature. The resist that is applied to the wafer can be at yet a third temperature. Resist thickness and uniformity are functions of these temperatures,[14,15] though the magnitudes of sensitivities also depend on the particular resist solvent. For example, reported values for changes in mean resist thickness range as low as 15 Å,[16] and as high as 65 Å[17] to 153 Å,[51] for a degree change in the temperature of the ambient air. The temperature of the chuck, and the materials it is composed of, can affect resist thickness uniformity.[18] Temperatures for the coating processes are usually chosen to minimize thickness variation across wafers and wafer-to-wafer, while the process average is modulated through the spin speed and resist viscosity.

Parameters which affect resist thickness and uniformity	
Temperatures	Resist
	Ambient air
	Wafer
	Chuck
	Softbake hotplate
Spin speeds	Speed during dispense
	Final spin speed
	Accelerations
Dispense	Resist volume
	Dispense rate
	Arm movement
Timing	Casting time
	End of dispense to cast delay
Exhaust	Velocity

Table 4.2. The parameters that affect resist thickness and uniformity.

Besides temperature, there are a number of other parameters that determine resist thickness uniformity and reproducibility. Humidity will also affect resist thickness. Sensitivities of 11 – 25 Å for a 1% change in relative humidity have been reported.[27, 19] Sensitivity to humidity is very dependent upon the resist solvent.

Exhaust is a critical parameter, which needs to be controlled dynamically,[20] since the exhaust provided to most semiconductor equipment will fluctuate as a consequence of varying loads from other equipment in the wafer fab. Equipment designers have recently paid closer attention to exhaust,[21, 22] which also can affect defect levels significantly as well as resist thickness.

For a given track and resist dispense system, there is a dispense volume which optimizes resist uniformity.[23] To minimize resist costs, this optimum volume should be small for well-designed equipment. Typical dispense volumes for coating 200 mm wafers range from 1 to 5 cc per wafer. This volume is very sensitive to the surface material composition and preparation, as well as the capability provided by the coating hardware. Resist spin speed and dispense arm movement are other parameters which affect the across-wafer resist thickness uniformity.[24, 25]

4.2.3 Focus

Focus is controlled by the wafer stepper. The separation between the lens and the wafer is measured prior to each exposure or during each scan, and the wafer height is adjusted with the objective of maintaining a constant lens-to-wafer distance. (On some older steppers it was the lens height which was changed.) The tilt of the exposure area relative to the optical axis can be determined by measuring the lens-to-wafer separation at several points on the wafer. There are a number of factors that influence the degree of focus control, many of which relate to the basic mechanism of the focus system. Three methods for focusing wafers are used on steppers:

- Optical.
- Capacitance.
- Pressure.

The optical method, illustrated in Fig. 4.10(a) is the technique used most commonly. In this approach, light is reflected from the substrate with a glancing angle of incidence. The reflected light will hit the detector at different positions, depending upon the vertical position of the substrate. In Fig. 4.10(a), light which reflects off a wafer at the position indicated by the solid line will hit the detector at a different position than light which reflects from a wafer at the position indicated by the dashed line. Detectors that can distinguish the position at which the light strikes the detector will be able to measure focus. The degree to which the optical method is insensitive to substrate film type and topography is critically dependent upon the system design.[26] For example, the optical system may detect the most strongly reflecting surface—the position of the resist film will be found at a different distance from the lens if it is sitting on top of a thick oxide film versus a bare silicon substrate. Process engineers must determine empirically if their combination of steppers and substrates creates such a sensitivity. Metal and thick oxide films are the substrates most commonly susceptible to focus errors. Minimizing the angle θ and using multiple wavelengths will provide the least sensitivity to films on the wafer surface.[27, 28, 29]

Two other methods of locating the wafer surface are shown in Fig. 4.10. In one method pressurized air is forced into a block which contains a pressure sensor. The pressure in the block will depend upon the gap between the block and the wafer surface. By mounting such blocks on the sides of lenses, the separations between lenses and wafers can be measured. This method is sensitive only to the top surface of the resist film and is independent of substrate composition. It does require calibration to the ambient barometric pressure, and this type of focus sensor cannot measure the height of the wafer surface directly below the lens during exposure, since the sensor would interfere with the imaging.

Capacitance sensors are also used to measure the lens-to-wafer distance. A capacitor is created where one plate is the wafer [Fig. 4.10(c)]. Capacitance C is given by:

$$C = \frac{\varepsilon A}{d} \tag{4.4}$$

where A is the area of the capacitor, d is the separation between the plates, and ε is the dielectric constant of the material between the plate. The separation between the lens and wafer can be determined by a measurement of capacitance

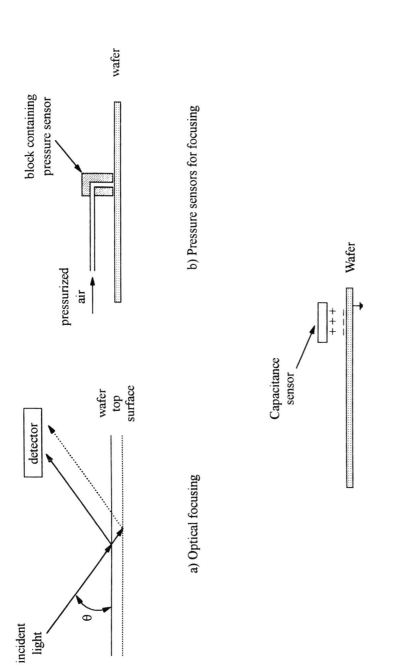

a) Optical focusing

b) Pressure sensors for focusing

c) Capacitance gauges for focusing

Figure 4.10. Different methods for measuring focus.

because of its dependence on d. Like the optical method, this technique is sensitive to the composition of the films on the silicon wafers.

All focus systems need calibration, and focus metrology has long been a problem. For many years best focus has been determined by patterning focus and exposure matrices. Inspection using optical microscopes was then used to identify best focus. Various patterns were used, including minimum lines and checkerboard patterns.[30] These methods suffer from inaccuracy and degrees of subjectivity, causing variation among operators. For linewidths > 0.5 µm optical linewidth measurement systems can be used to eliminate the subjective element.[31] As linewidths have fallen below 0.25 µm, optical inspection and linewidth measurement systems have become inadequate. Scanning electron microscopes can be used, but these systems are typically very slow in processing the large numbers of images required to determine focus accurately. Two methods have been introduced which provide greater accuracy and speed for measuring focus: phase-shifting focus and stepper self-metrology.

The phase-shifting focus monitor is based upon asymmetries in imaging which occur when patterns with 90° phase shifters are defocused.[32,33] Consider the pattern shown in Fig. 4.11. On the mask there is a chrome line, with 0° phase shifting on one side and 90° phase shifting on the other side. The aerial image of the line is shown in Fig. 4.12, at best focus, and out of focus for an ideal lens. At best focus, the aerial image is symmetric, but when defocused, the image becomes asymmetric. More importantly, the asymmetry moves in opposite directions, depending upon whether the wafer is moved closer or further from the lens. When patterning wafers, the resulting resist feature will move as a consequence of this asymmetry. Equipment that is normally used for measuring overlay can be used to measure defocus by the use of the structure shown in Fig. 4.13, for which defocus will cause effective shifts in overlay. This monitor has demonstrated the ability to measure focus to a precision of 50 nm.

The image shifts shown in Fig. 4.12 represent ideal optics. In practice, centering of the structures in Fig. 4.13 does not always represent best focus, and the calibration of the monitor has been found to be dependent on the numerical aperture and partial coherence settings of the stepper.[34] Nevertheless, the phase-shifting focus monitor can be calibrated, and thereafter provide a high precision (50 nm) tool for measuring defocus. Once calibrated, this monitor can be used to measure focus at many points across wafers, and the data can be collected quickly using the tools normally used for automatically measuring overlay. This capability has enabled focus problems to be identified, such as poor chucking.[35]

The determination of a stepper's best focus setting by processing wafers on the stepper and measuring the wafers on scanning electron microscopes or overlay measurement tools is time consuming and requires coordination of the processing equipment and metrology tools. More efficient and operationally

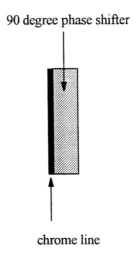

90 degree phase shifter

chrome line

Figure 4.11. A chrome line with different phase light transmitted on its two sides.

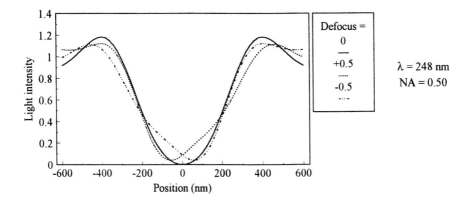

Figure 4.12. The light intensity profile of the line shown in Fig. 4.11, for different defocus.

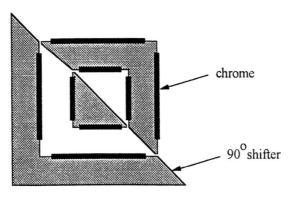

Figure 4.13. An overlay measurement structure which can be used to measure defocus.

simpler is stepper self-metrology,[36, 37, 38, 39] where steppers can determine their own best focus setting. In some situations patterns are imaged into resist,[40] while other self-metrology systems do not require wafers at all.

The glasses used for i-line lenses absorb small amounts of light. As a consequence the lenses heat, resulting in shifts of focus.[41] Modern steppers will automatically shift focus as wafers are exposed and the lenses are heated. The amount of heating that takes place is a function of the quantity of light which is transmitted through the reticle and into the lens, which is a function of pattern density, types of features, illumination, and the numerical aperture of the projection optics. Steppers often have sensors on the stage that can be used to measure the amount of transmitted light directly. As steppers sit idle the lenses will cool, and the stepper software must have the correct time constants to account properly for lens cooling as well as heating.

Focus is also a function of barometric pressure.[2, 42] Modern lenses are either pressurized to maintain a constant pressure inside, or the stepper software is used to adjust focus as barometric pressure varies. The software parameters required tend to vary from lens to lens, even for lenses of the same design. These parameters do not require changing, but software is always susceptible to corruption, and focus errors can result from improper compensation for barometric pressure, even for systems initially set up correctly. Good control of steppers, and all other equipment, requires good records and back-up of software parameters.

4.2.4 Bake temperatures

Much of resist processing involves thermal processes and chemical reactions. Reaction rates are typically temperature dependent:

$$\text{rate} = A_o e^{-\frac{E_a}{kT}}, \tag{4.5}$$

where k is Boltzmann's constant, T is the absolute temperature, E_a is an activation energy, and A_o is a rate constant. The specific requirements for control depend upon the particular resist materials, which determine the rate constants, but generally it is important to control the temperatures used for processing photoresist.

Bake times also need to be controlled. Consider the dose-to-clear data shown in Fig. 4.14.[43] A one second change in a 20 second bake changes E_0 by 1.2%, and changes E_0 by only 0.5% for a 60 second bake. The bake time should be long enough so that the time over which the bake can be controlled does not change the process significantly.

Effective bake times are determined by more than the length of time wafers are on hotplates. Once wafers are removed from hotplates they will begin to cool, and this cooling needs to occur in a controlled way. Consider the data shown in Fig. 4.15, which show the temperatures of wafers as they are placed on a hotplate, removed, and finally set on a chillplate.[44] When the wafers are placed at time = 0 on the hotplate their temperatures rise quickly to the hotplate temperature of 105° C. After 60 secs on the hotplate the wafers are lifted above the plate on pins, so that the wafers can then be picked up by a wafer transfer arm and moved to a chill plate. The temperature of the wafers declines slowly while the wafers are sitting on the pins. Because the wafer transfer arm may or may not be occupied moving other wafers, the wafers may sit on the pins for variable times. The change in critical dimensions was measured for a 10 second variation in times on the pins and was found to be 7.4 nm when the relative humidity was 42% and 10.8 nm when the relative humidity was 10.8% for a conventional g-line resist.[44] Chillplates were introduced to semiconductor resist processing in order to provide reproducible bake processes,[45] and this example shows that even subtle aspects of the bakes need to be considered in order to squeeze out the last few nanometers of process control.

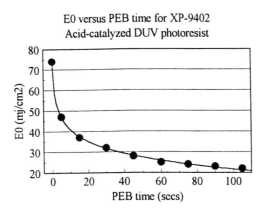

Figure 4.14. Changes in dose-to-clear as a function of post-exposure bake (PEB) time.

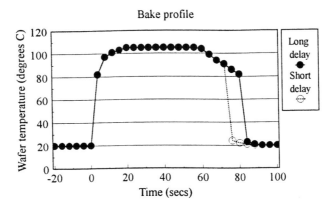

Figure 4.15. Changes in effective baking as a consequence of variable times between bake and chill.

The hotplate needs to be well designed in order to ensure good process control. The hotplate receives a thermal shock when a wafer is first placed on it, and the heating elements will respond. If the hotplate controller is not well designed, there may be a temperature overshoot before the wafer temperature is returned to the desired value. This overshoot can cause statistically measurable shifts in linewidth.[46] In order to have a well-controlled process, the wafer temperature needs to be controlled across each wafer, wafer-to-wafer, and during the entire heating and cooling cycle for each wafer.

In order to verify that bake temperature is well controlled, wafer temperatures need to be measured. This is accomplished typically by attaching temperature sensors to many points on a wafer and placing the wafer on a hotplate. Temperature readings for each sensor can be sampled in time, and a history of the wafer temperature can be collected. This method requires that accurate temperature sensors be used. Today, these thermal probes are usually platinum resistance temperature detectors (RTDs), with precision of $\pm 0.05^{\circ}C$ and an absolute accuracy of $\pm 0.1^{\circ}C$.[47] Use of wafers covered by temperature sensors is not straightforward (because of the delicate wires which run between the probes and the associated electronics), but can nevertheless be a very useful diagnostic tool.

4.2.5 Resist development

Resist development also involves a chemical reaction. Parameters that are expected to be critical are temperature and developer concentration. The effect of developer temperature on linewidth control was studied extensively in the early and mid-1980s.[48] During that time it became recognized that developer temperature can affect linewidths significantly, and good temperature control was implemented on the equipment used for developing wafers. Puddle develop on a track system is the method used most commonly today. The tubes carrying the developer for the wafer are jacketed, and temperature is controlled actively to within $\pm 0.2^{\circ}C$. Once the developer is on the wafer it will begin to cool because of evaporation. Because the rate of evaporation is greatest on the outside of the wafer, thermal gradients are generated, leading to different rates of development from the center to the edge of the wafer [tetramethyl ammonium hydroxide (TMAH) developers become more aggressive at lower temperatures]. Evaporation will also cause developer concentration to increase. These effects can be minimized by using developer that is dispensed at a low temperature to begin with and by having a low air flow around the wafer while the resist is developing.[49]

Developer composition also needs to be tightly controlled. The most common resists used for semiconductor lithography use aqueous basic developers. Early developers used sodium hydroxide or potassium hydroxide for the base. Because sodium and potassium can lead to reliability failures in MOS transistors, the most commonly used developer today is TMAH. To improve wetting of resist surfaces, surfactants are often added to TMAH developers, and buffers have been used to maintain constant developer pH of metal-ion developers during the development process. The chemical activity of multi-component developers is characterized by multiple inflection points[50] (Fig. 4.16). For this reason, many lithographers prefer to use developers which do not contain surfactants or buffers, since only a single component is then of concern, and the composition is easier to control. Developer is produced in batches, and the developer composition is very uniform within the batch. The introduction of a new batch of developer is a special event, and changes in linewidths can be correlated to transitions between batches. If no measurable changes in linewidths are observed when developer batches are changed, then the developer composition is being controlled adequately.

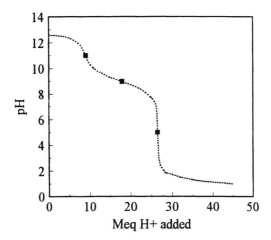

Figure 4.16. Titration curve for developer.

4.2.6 Humidity

Water is a required reactant in the chemistry of napthaquinone diazide photoresists, i.e., nearly all positive g- and i-line resists. It might therefore be expected that linewidth control depends on the relative humidity. In one experiment, linewidths were measured over a range of 30 – 60% relative humidity. Changes in linewidth were observed, as a function of relative humidity in the resist coater (Fig. 4.17), but after pursuing the problem further, it was found that the dominant effect was a change in resist thickness[51] (Fig. 4.18). The linewidths changed because of thin film optical effects.[52] In another experiment, the humidity was controlled in the exposure tool, and the resulting level of photoacid was measured in several resists using Fourier transform infrared spectroscopy.[53] A typical result from that work is shown in Fig. 4.19. While water is required for the chemical reaction in napthaquinone diazide photoresists, it appears that the photochemistry will be stable as long as the relative humidity does not fall below 35%. In order to control resist thickness, however, the humidity must be controlled. For DUV resists that do not require water as part of their chemistries, this latter fact remains significant.

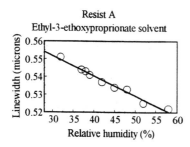

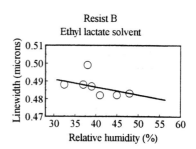

Figure 4.17. Changes in linewidth found as a fuction of humidity. The change in linewidth is 1.1 nm per 1% humidity for Resist A and 0.4 nm per 1% humidity for Resist B.

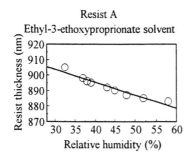

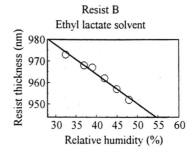

Figure 4.18. Resist thickness as a function of humidity. The thickness of Resist A changes by 0.8 nm per 1% of humidity, and the thickness of Resist B changes 1.1 nm for every 1% change in relative humidity.

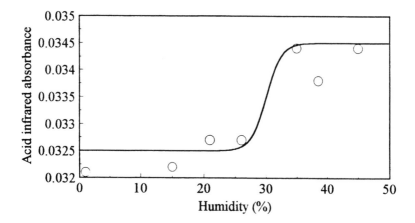

Figure 4.19. Infrared absorbance of the acid produced by exposure of a novolac/diazonapthaquinone resist.

4.2.7 DUV resists - special considerations

Acid catalyzed, chemically amplified resists are susceptible to poisoning by part-per-billion levels of ambient amines.[54,55] The photochemistry of these resists involves the generation of an acid upon exposure to DUV light. Amines that diffuse into the resist film can neutralize these photoacids and prevent deprotection from taking place. The resist will not fully develop where this occurs. One of the most common sources of amine contamination is the ambient air, so the top of the resist is where the photoacid is first neutralized. A typical result of this poisoning of the resist chemistry is shown in Fig. 4.20. For slight poisoning the photoresist features are slightly flared at the top, but the resist pattern can be completely bridged following longer exposures to ambient amines. Contamination at lower levels may not have such dramatic consequences, but there will be significant changes in the dimensions of critical linewidths.[56]

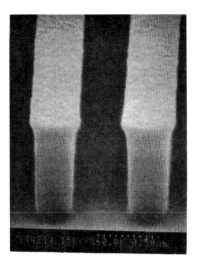 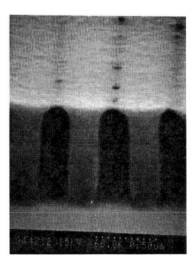

Figure 4.20. Apex E poisoned by ambient amines. The example on the left is slight poisoning, while the example on the right is the result of longer exposure to amines.

Special precautions need to be taken in order to prevent, or at least minimize, the consequences of resist poisoning. The most common method of reducing the level of ambient amines in the air around resist-covered wafers is filtration. Activated charcoal is used to remove complex organic amines, such as N-methyl-pyrilidone, while weak acids, such as citric acid, are used to neutralize ammonia. The air in the steppers and resist process equipment is often recirculated to maximize the removal of amines.

Substrates can also cause difficulties with chemically amplified resists, including titanium nitride, silicon nitride, and various forms of silicon dioxide.[57,58,59] These substrates can cause footing at the bottom of resist profiles. Processes should be set up initially to be reasonably free of such footing, but resist poisoning by substrates is sensitive to deposition conditions. If the deposition of substrate films changes, the lithography process may be affected. Under such circumstances the linewidth monitors described in the previous chapter are invaluable, providing independent tests that the lithography process has remained stable. Sensitive monitors of the deposition process can also prove useful. Substrates may also be treated to reduce the extent of resist poisoning.[60]

Chemically amplified resists are susceptible to poisoning by amines between the time of exposure, when the photoacids are generated, and post-exposure bake, which is when the deprotection reaction takes place. It is essential to minimize this time between exposure and bake in order to keep the effects of poisoning to a minimum. Moreover, it is important to maintain consistent times between exposure and bake. This can be accomplished through the design and programming of resist processing equipment. While controlling the time between exposure and the post-exposure bake is of particular importance for DUV resists, it should be noted that similar sensitivities have been observed for novolac resists,[61] though as a consequence of a different mechanism.

The newest generations of DUV resists have been designed to reduce this poisoning effect.[62] Processes that use these resists are easier to control. Sensitivity to amines is a significant parameter when choosing a DUV photoresist, and this factor needs to be considered during resist evaluations.

4.2.8 Contributions from reticles

Reticles represent sources of variation in a manufacturing facility or pilot line, because the average dimension and the pattern of variation changes from reticle to reticle. The consequence of this was observed in a facility producing application-specific integrated circuits, where there were multiple new products introduced every month, and it was found that the reticles were the largest source of linewidth variation [Ch. 3, Ref. 6]. As linewidths shrink, a point is reached where the dimensions on the wafer change faster than on the reticle, with appropriate scaling for lens reduction. Consider the data in Fig. 4.21.[63] For features larger than 300 nm the image is transferred from the reticle to the wafer at a ratio of 4:1, the nominal reduction of the 0.5 NA Micrascan II lens. Consequently, deviations in size for features on the reticle are reduced in magnitude by a factor of four. For small features, this error reduction decreases to a factor of only two. Processes that operate at or near the diffraction limit of the optics will have smaller mask error reduction factors than those which are far from the diffraction limit. For this reason, the nominal reduction factor of 4× for the lenses used for patterning the most critical layers is being reconsidered by lithographers.

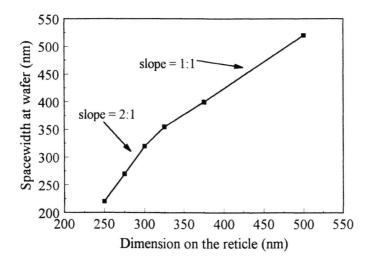

Figure 4.21.   Wafer dimension versus reticle feature size, (scaled for the 4× reduction of the lens). The imaging system had a numerical aperture of 0.5 and a wavelength of 248 nm.

4.3 MAXIMIZING THE PROCESS WINDOW

The best process is one that varies the least for the normal variations in the equipment and materials that are used. The most effective techniques for determining the optimum process are referred to as "design of experiments," which include factorial designs,[64] surface response techniques,[65] and Taguchi methods.[66] Each of these techniques is legitimately a topic for an entire book and is therefore beyond the scope of this text. The power of these methods lies in the ability to efficiently identify sources of variation and process optimums without a detailed understanding of the physical or chemical mechanisms driving the variation.

The purpose of optimization in the context of process control is the minimization of variation. As discussed in Chapter 1, many wafers are required to determine the level of variation, which is not consistent with efficiency. It is more effective to identify a parameter whose mean must be either maximized or minimized. Consequently, in the context of linewidth control, most optimization programs maximize the focus-exposure window.[67] There are a number of reasons why this is an effective approach to minimizing linewidth variation. First, focus is a critical parameter for controlling sub-micron linewidths, particularly those smaller than 0.5 μm. The maximization of focus-exposure windows therefore includes this critical parameter. Second, maximizing exposure latitude addresses a number of problems. As discussed earlier in this chapter, even though steppers control exposure doses reasonably well, there are many phenomena, such as thin film optical effects, which modulate the amount of actinic light coupled into photoresist films. Large exposure latitude reduces sensitivity to these effects. Finally, focus-exposure matrices can be patterned on single wafers, enhancing the efficiency of this approach. Modern steppers have variable numerical aperture and illumination. Process optimization must include the resist process as well as these stepper parameters.

It has been observed by the author of this text and others[68, 69, 70] that process latitude can often be maximized when the dose to replicate mask features at their nominal dimensions (E_{1-1}) is much larger than the dose-to-clear (E_0):

$$\frac{E_{1-1}}{E_0} \gg 1, \qquad (4.6)$$

particularly on reflective substrates. (A different process optimum may occur on substrates with low reflectivity.) For example, if the reticle features consist of 250 nm lines and spaces (times the reduction factor of the lens), then E_{1-1} is the exposure dose which produces 250 nm lines and spaces in the resist. This dose is typically larger than the minimum dose required to clear a large exposed area (> 1 μm on a side) of positive photoresist.

An overview of the physics and chemistry relevant to the formation of patterns in photoresist was presented in this chapter. Such scientific understanding is necessary for the efficient optimization and control of linewidths. The technical aspects of overlay are the subjects of the next chapter.

---

[1] K. Ishikawa, *Guide to Quality Control*, Asian Productivity Organization, White Plains, New York (1991).

[2] H. J. Levinson and W. H. Arnold, "Focus: The Critical Parameter for Sub-micron Lithography," *J. Vac. Sci. Technol.*, B5(1), pp. 293 – 298 (1987).

[3] IMEC is the acronym for Interuniversity Microelectronics Centre, Leuven, Belgium.

[4] K. Ronse, R. Pforr, L. van den Hove, and M. Op de Beeck, "CD Control: the Limiting Factor for *i*-line and Deep-UV Lithography?" OCG Microelectronics Seminar, pp. 241 – 254 (1995).

[5] K. Ronse, M. Op de Beeck, A. Yen, K-H. Kim, and L. van den Hove, "Characterization and Optimization of CD Control for 0.25 μm CMOS Applications," SPIE Vol. 2726, pp. 555 – 563 (1996).

[6] P. Schoenborn and N. F. Pasch, "Process Sensitivity Analysis: Applications to Photolithography," SPIE Vol. 1087, pp. 290 – 298 (1989).

[7] Z. Krivokapic, W. D. Heavlin, and D. Kyser, "Process Capabilities of Critical Dimensions at Gate Mask," SPIE Vol. 2440, pp. 480 – 491 (1995).

[8] *Guide for Reducing Quality Costs*, 2nd Edition. ASQ Press, Milwaukee (1987).

[9] Finle Technologies, Austin, Texas.

[10] R. W. Leonhardt and T. R. Scott, "Deep-UV Excimer Laser Measurements at NIST," SPIE Vol. 2439, pp. 448 – 459 (1995).

[11] K. H. Kim, W. S. Han, C. H. Kim, H. Y. Kang, C. G. Park, and Y. B. Koh, "Characteristics of Standing Wave Effect of Off-axis Illumination Depending on two Different Resist Systems and the Polarization Effect of Stepper," SPIE Vol. 2197, pp. 42 – 53 (1994).

[12] M. Born and E. Wolf, *Principles of Optics*, 6th Edition, Pergamon Press, New York (1990).

[13] J. Sturtevant and B. Roman, "Antireflection Strategies for Advanced Photolithography," *Microlithography World*, pp. 13 – 21 (1995).

[14] G. MacBeth, "Thermal Effects in Photoresist Coating Processes," KTI Microelectronics Seminar, pp. 327 – 340 (1988).

[15] S. Dick and B. Greenstein, "Improved Photolithography Process Performance Through the Use of an Integrated Photosector," KTI Microelectronics Seminar, pp. 1 – 8 (1989).

[16] M. Reihani, "Environmental Effects on Resist Thickness Uniformity," *Semiconductor International*, pp. 120 – 121 (June, 1992).

[17] T. Batchelder, M. Ha, R. Haney, and W. Lee, "Sub-micron Linewidth Control with Automatic Optical Monitors," KTI Microelectronics Seminar, pp. 231 – 242 (1989).

[18] D. P. Birnie III, B. J. J. Zelinski, S. P. Marvel, S. M. Melpolder, and R. L. Roncone, "Film/Substrate/Vacuum-Chuck Interactions During Spin-Coating," *Optical Engineering*, Vol. 31(9), pp. 2012 – 2020 (1992).

[19] M. Watts and S. Williams, "A Novel Method for the Prediction of Process Sensitivity in Photolithography," SPIE Vol. 1261, pp. 345 – 359 (1990).

120992272629892128122992

[20] R. Murray, P. T. Edwin, V. Taburaza, and J. Olin, "Airflow Controller Improves Photoresist Spin/Coat Uniformity," *Semiconductor International*, pp. 172 –173 (April, 1987).

[21] L. Matter, J. Zook, M. Hinz, J. Banas, and S. Ibrani, "New Coat Bowl Design Improves Photoresist Uniformity and Decreases Particle Contamination," Olin Microlithography Seminar Poster Session (1997).

[22] X. Zhu, F. Liang, A. Haji-Sheikh, and N. Ghariban, "A Computational and Experimental Study of Spin Coater Air Flow," SPIE Vol. 3333, pp. 1441 – 1451 (1998).

[23] W. J. Daughton, P. O'Hagan, and F. L. Givens, "Thickness Variance of Spun-On Photoresist, Revisited," Proceedings of the Kodak Microelectronics Seminar, pp. 15 – 20 (1978).

[24] K. Kemp, D. Williams, J. Daggett, J. Cayton, S. Slonaker, and R. Elliott, "Critical Dimension Performance Characterization of an Advanced DUV Process Cell," Olin Microlithography Seminar, pp. 99 – 108 (1996).

[25] D. Boutin, A. Blash, J. P. Caire, D. Poncet, P. Fanton, M. Danielou, and B. Previtali, "Resist Coating Optimization on Eight Inches Deep UV Litho Cell Modelization and Application to 0.25 μm Technology, SPIE Vol. 2439, pp. 495 – 502 (1995).

[26] T. O. Herndon, C. E. Woodward, K. H. Konkle, and J. I. Raffel, "Photocomposition and DSW Autofocus Correction for Wafer-Scale Lithography," Proceedings of the Kodak Microelectronics Seminar, pp. 118 – 123 (1983).

[27] A. Suzuki, S. Yabu, and M. Ookubo, "Intelligent Optical System for a New Stepper," SPIE Vol. 772, pp. 58 – 65 (1987).

[28] J. E. van den Werf, "Optical Focus and Level Sensor for Wafer Steppers," *J. Vac. Sci. Technol.*, Vol. 10(2), pp. 735 – 740 (1992).

[29] M. A. van den Brink, J. M. D. Stoeldraijer, and H. F. D. Linders, "Overlay and Field by Field Leveling in Wafer Steppers Using an Advanced Metrology System," SPIE Vol. 1673, pp. 330 – 344 (1992).

[30] J. W. Gemminck, "Simple and Calibratable Method for Determining Optimal Focus," SPIE Vol. 1088, pp. 220 – 230 (1989).

[31] S. Venkataram, C. Olejnik, G. Flores, and D. Tien, "An Automated Technique for Optimizing Stepper Focus Control," SPIE Vol. 2725, pp. 765 – 778 (1996).

[32] T. A. Brunner, A. L. Martin, R. M. Martino, C. P. Ausschnitt, T. H. Newman, and M. S. Hibbs, "Quantitative Stepper Metrology Using the Focus Monitor Test Mask," SPIE Vol. 2197, pp. 542 – 549 (1994).

[33] R. D. Mih, A. Martin, T. Brunner, D. Long, and D. Brown, "Using the Focus Monitor Test Mask to Characterize Lithographic Performance," SPIE Vol. 2440, pp. 657 – 666 (1995).

[34] E. R. Sherman and C. Harker, "Characterization and Monitoring of Variable NA and Variable Coherence Capable Photo Steppers Utilizing the Phase Shift Focus Monitor Reticle," SPIE Vol. 2439, pp. 61 – 69 (1995).

[35] T. A. Brunner and R. D. Mih, "Simulations and Experiments with the Phase Shift Focus Monitor," SPIE Vol. 2726, pp. 236 – 243 (1996).

[36] T. A. Brunner and S. M. Stuber, "Characterization and Setup Techniques for a 5X Stepper," SPIE Vol. 633, pp. 106 – 112 (1986).

[37] M. van den Brink, H. Franken, S. Wittekoek, and T. Fahner, "Automatic On-Line Wafer Stepper Calibration System," SPIE Vol. 1261, pp. 298 – 314 (1990).

[38] T. A. Brunner, S. Cheng, and A. E. Norton, "Stepper Image Monitor for Precise Setup and Characterization," SPIE Vol. 922, pp. 366 – 375 (1988).

[39] R. Pforr, S. Wittekoek, R. van den Bosch, L. van den Hove, R. Jonckheere, T. Fahner, and R. Seltmann, "In-process Image Detecting Technique for Determination of Overlay and Image Quality for the ASM-L Wafer Stepper," SPIE Vol. 1674, pp. 594 – 609 (1992).

[40] P. Dirksen, W. de Laat, and H. Megens, "Latent Image Metrology for Production Wafer Steppers," SPIE Vol. 2440, pp. 701 – 711 (1995).

[41] T. A. Brunner, J. M. Lewis, and M. P. Manny, "Stepper Self-Metrology Using Automated Techniques," SPIE Vol. 1261, pp. 286 – 297 (1990).

[42] K. Hale and P. Luehrman, "Consistent Image Quality in a High Performance Stepper Environment," Proceedings of the Kodak Microelectronics Seminar, pp. 29 – 46 (1986).

[43] J. S. Petersen, C. A. Mack, J. W. Thackeray, R. Sinta, T. H. Fedynyshyn, J. J. Mori, J. D. Byers, and D. A. Miller, "Characterization and Modeling of a Positive Acting Chemically Amplified Resist," SPIE Vol. 2438, pp. 153 – 166 (1995).

[44] J. M. Kulp, "CD Shift Resulting from Handling Time Variation in the Track Coat Process," SPIE Vol. 1466, pp. 630 – 640 (1991).

[45] G. MacBeth, "Prebaking Positive Photoresists," Proceedings of the Kodak Microelectronics Seminar, pp. 87 – 92 (1982).

[46] O. D. Crisalle, C. L. Bickerstaff, D. E. Seborg, and D. A. Mellichamp, "Improvements in Photolithography Performance by Controlled Baking," SPIE Vol. 921, pp. 317 – 325 (1988).

[47] J. Parker and W. Renken, "Temperature Metrology for CD Control in DUV Lithography," *Semicond. International*, pp. 111 – 116, (September 1997).

[48] C. M. Garza, C. R. Szmanda, and R. L Fischer, "Resist Dissolution Kinetics and Submicron Process Control," SPIE Vol. 920, pp 321 – 338 (1988).

[49] M. K. Templeton, J. B. Wickman, and R. L. Fischer, Jr., "Submicron Resolution Automated Track Development Processes, Part 1: Static Puddle Development," SPIE Vol. 921 pp. 360 – 372 (1988).

[50] W. D. Hinsberg and M. L. Gutierrez, "Effect of Developer Composition on Photoresist Performance," Proceedings of the Kodak Microelectronics Seminar, pp. 52 – 56 (1983).

[51] E. Bokelberg and W. Venet, "Effects of Relative-humidity Variation on Photoresist Processing," SPIE Vol. 2438, pp. 747 – 752 (1995).

[52] W. H. Arnold and H. J. Levinson, "High Resolution Optical Lithography Using an Optimized Single Layer Photoresist Process," Proceedings of the Kodak Microelectronics Seminar, pp. 80 – 92 (1983).

[53] J. A. Bruce, S. R. DuPuis, and H. Linde, "Effect of Humidity on Photoresist Performance," Proceedings of the OCG Microlithography Seminar, pp. 25 – 41 (1995).

[54] S. A. MacDonald, C. G. Wilson, and J. M. J. Fréchet, "Chemical Amplification in High-Resolution Imaging Systems," *Accounts of Chemical Research*, Vol. 27(6), pp. 151 – 158 (1994).

[55] K. R. Dean, D. A. Miller, R. A. Carpio, and J. S. Petersen, "Airborne Contamination of DUV Photoresists: Determining the New Limits of Processing Capability," Proceedings of the Olin Microlithography Seminar, pp. 109 – 125 (1996).

[56] K. van Ingen Schenau, M. Reuhman, and S. Slonaker, "Investigation of DUV Process Variables Impacting Sub-Quarter Micron Imaging," Proceedings of the Olin Microlithography Seminar, pp. 63 – 80 (1996).

[57] N. Thane and G. A. Barnes, "Process Compatibility of Anti-reflection Coatings with Various Deep UV Resists," Proceedings of the OCG Microlithography Seminar, pp. 61 – 76 (1993).

[58] J. S. Sturtevant, S. Holmes, S. Knight, D. Poley, P. Rabidoux, L. Somerville, T. McDevitt, A. Stamper, E. Valentine, W. Conley, A. Katnani, and J. Fahey, "Substrate Contamination Effects in the Processing of Chemically Amplified DUV Photoresists," SPIE Vol. 2197, pp. 770 – 780 (1994).

[59] K. R. Dean, R. A. Carpio, and G. K. Rich, "Investigation of Deep Ultraviolet Photoresists on TiN Substrates," SPIE Vol. 2438, pp. 514 – 528 (1995).

[60] A. Usujima, K. Tago, A. Oikawa, and K. Nakagawa, "Effects of Substrate Treatment in Positive Chemically-Amplified Resist," SPIE Vol. 2438, pp. 529 – 539 (1995).

[61] Y-K. Hsiao, C-H. Lee, S-L. Pan, K-L. Lu, and J-C. Yang, "A Study of the Relation Between Photoresist Thermal Property and Wettability," Proceedings of the Olin Microlithography Seminar, pp. 315 – 323 (1996).

[62] H. Ito, W. P. England, H. J. Clecak, G. Breta, H. Lee, D. Y. Yoon, R. Sooriyakumaran and W. D. Hinsberg, "Molecular Design for Stabilization of Chemical Amplification Resist Toward Airborne Contamination," SPIE Vol. 1925, pp. 65 – 75 (1993).

[63] W. Maurer, K. Satoh, D. Samuels, and T. Fischer, "Pattern Transfer at k_1 = 0.5: Get 0.25 μm Lithography Ready for Manufacturing," SPIE Vol. 2726, pp. 113 – 124 (1996).

[64] G. E. P. Box, W. G. Hunter, and J. S. Hunter, *Statistics for Experimenters*, John Wiley & Sons, New York (1978).

[65] G. E. P. Box and N. R. Draper, *Empirical Model-building and Response Surfaces*, John Wiley & Sons, New York (1987).

[66] G. Taguchi, *System of Experimental Design, Vols. 1 and 2*, American Supplier Institute Press, Dearborn, Michigan (1987).

[67] For example, K. Petrillo, "Process Optimization of Apex-E," SPIE Vol. 1926, pp. 176 – 187 (1993).

[68] S. G. Hansen, G. Dao, H. Gaw, Q-D. Qian, P. Spragg, and R. J. Hurditch, "Study of the Relationship Between Exposure Margin and Photolithographic Process Latitude and Mask Linearity," SPIE Vol. 1463, pp. 230 – 244 (1991).

[69] J. S. Petersen and A. E. Kozlowski, "Optical Performance and Process Characterization of Several High Contrast Metal-Ion Free Developer Processes," SPIE Vol. 469, pp. 46 – 56 (1984).

[70] J. S. Petersen, "An Experimental Determination of the Optical Lithographic Requirements for Sub-Micron Projection Printing," SPIE Vol. 188, pp. 540 – 567 (1989).

CHAPTER 5
OVERLAY

5.1 OVERLAY MODELS

A discussion of overlay should begin with definitions. Overlay is defined in SEMI Standard P18-92, "Specifications for Overlay Capabilities of Wafer Steppers," as[1]:

> Overlay - A vector quantity defined at every point on the wafer. It is the difference, $\vec{O}$, between the vector position, $\vec{P}_1$, of a substrate geometry and the vector position of the corresponding point, $\vec{P}_2$, in an overlaying pattern, which may consist of photoresist:

$$\vec{O} = \vec{P}_2 - \vec{P}_1. \qquad (5.1)$$

A related quantity, registration, is defined similarly:

> Registration - A vector quantity defined at every point on the wafer. It is the difference, $\vec{R}$, between the vector position, $\vec{P}_1$, of a substrate geometry and the vector position of the corresponding point, $\vec{P}_0$, in a reference grid:

$$\vec{R} = \vec{P}_1 - \vec{P}_0. \qquad (5.2)$$

Note that overlay is a relative quantity, while registration is an error compared to an absolute standard, $\vec{P}_0$.

Overlay errors can be considered in terms of a hierarchy (Fig. 5.1). Most fundamental are those errors which occur when only a single stepper and ideal substrates are used, and where the latter provide high signal-to-noise alignment signals. This basic set of overlay errors is well described by overlay models, which will be discussed in detail shortly. When more than a single stepper is used, an additional set of overlay errors is introduced, referred to as matching errors. Finally, there are process-specific contributions to overlay that can result in non-ideal alignment targets. A full accounting of overlay errors in terms of a hierarchy is useful for assessing and improving overlay.[2]

The control of overlay requires the use of models, because these models are employed directly in the software of wafer steppers when aligning wafers, and the model parameters are changed when adjustments are made to wafer steppers. Overlay models conform to the physics of wafer steppers. There are two categories of overlay errors: intrafield, which involve the variation of overlay errors within exposure fields, and interfield errors, which are the errors which vary from field to field across wafers.

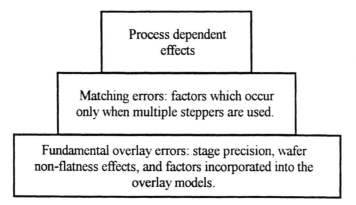

Figure 5.1. The hierarchy of overlay errors. The foundation includes those factors common to all situations.

Interfield errors can be understood by considering the overlay error at equivalent points in each exposure field. Conventionally this is the center of the exposure field. Let (X,Y) be the coordinate of a point on the wafer corresponding to the center of some exposure field. The center of the wafer is the most convenient location for the origin of the coordinate system. The overlay error at the point (X,Y) is given by:

$$\Delta X = f_X(X,Y) \tag{5.3}$$

$$\Delta Y = f_Y(X,Y) \tag{5.4}$$

where ΔX is the overlay error in the X direction and ΔY is the error in the Y direction. The overlay error ΔX can be expressed in a MacLaurin series of the function f_X :

$$\Delta X = T_X + E_X X - R_X Y + \text{higher order terms} + e_X . \tag{5.5}$$

The reason for the negative sign prior to the $R_X Y$ term will be discussed shortly. There is a similar expression for ΔY:

$$\Delta Y = T_Y + E_Y Y + R_Y X + \text{higher order terms} + e_Y . \tag{5.6}$$

In addition to the higher order terms there are residual errors, e_X and e_Y, at every measured point which do not conform to the model. The most common source of residual error is the imprecision of the stepper's stage, which is on the order of 15 nm (3σ) for the current generation of wafer steppers.[3] Another source of non-modeled error arises when multiple steppers are used. This is the issue of matching, which will be discussed in the next section.

The linear terms in Eqs. 5.5 and 5.6 have physical meanings. Each layer can be considered as patterns printed on a rectangular, or near-rectangular, grid (Fig. 5.2).

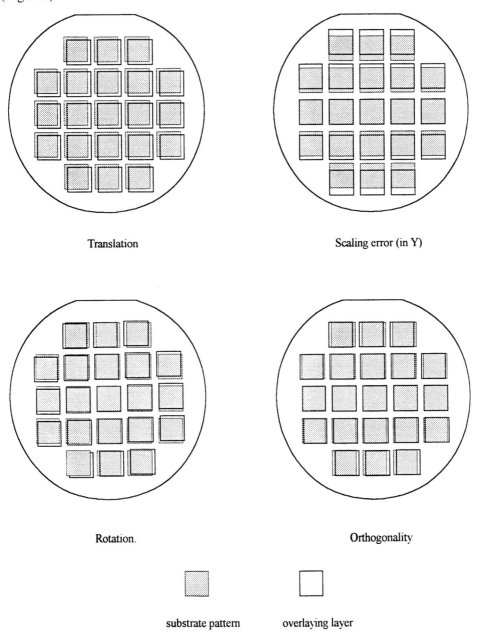

Figure 5.2. Interfield overlay errors.

The parameters T_X and T_Y represent translation errors in the X and Y directions, respectively, and indicate an overall shift of one grid relative to the other. The factors E_X and E_Y are scale errors, which represent the errors made by the stepper in compensating for wafer expansion or contraction. Scale errors are dimensionless and are usually expressed in parts-per-million (ppm). The coefficients R_X and R_Y are rotation factors. When $R_X = R_Y$ one grid is rotated relative to the other, and this accounts for the sign convention chosen in Eq. 5.5. If the angle between the axes of the grids are not equal, then $R_X \neq R_Y$. This latter error is referred to as an orthogonality error, since it represents the situation in which the grid for at least one of the layers (substrate or overlaying) has non-orthogonal axes. The rotation prefactors are tangents of angles, but since

$$\theta \approx \tan \theta \qquad (5.7)$$

for small angles θ, the rotations are usually expressed in radian measure, typically in units of microradians.

The type of alignment used with most wafer steppers is some form of global or enhanced global alignment, in which the positions in space of at least two points on the wafer are determined. When only two points are determined by the alignment process it is possible to compute four of the parameters in Eqs. 5.5 and 5.6. For such two point global alignment it is usually assumed that $E_X = E_Y$ and $R_X = R_Y$. By aligning the wafer at three or more (non-colinear) points on the wafer it is possible to estimate all of the linear terms in Eqs. 5.5 and 5.6. The overlaying pattern is adjusted according to these linear corrections, through the software of the stepper. This model is used directly in the stepper software and is relatively unchanged from the original interfield model introduced by Perloff.[4]

Some overlay errors cannot be detected during alignment. For example, it has been observed that wafers heat during i-line exposures, causing wafer scaling errors that arise after the wafers have been aligned.[5] Non-linear errors will not be corrected by the software of most steppers, which use a linear model. Some errors which are not included in the model of Eqs. 5.5 and 5.6 will be discussed in the next section.

The overlay errors may also vary across each exposure field. These types of errors are referred to as intrafield errors, examples of which are shown in Fig. 5.3. Consider, for example, a magnification error. The lens reduction typically has nominal values of 4:1 or 5:1, but the magnification will deviate from these nominal values by some small amount. When there is wafer expansion and contraction between masking steps, it is necessary for the stepper to be programmed to measure this change and compensate for it, not only in the grid terms, but in the size of each field as well. Errors in doing this correction result in magnification errors. The other primary intrafield errors are shown in Fig. 5.3.

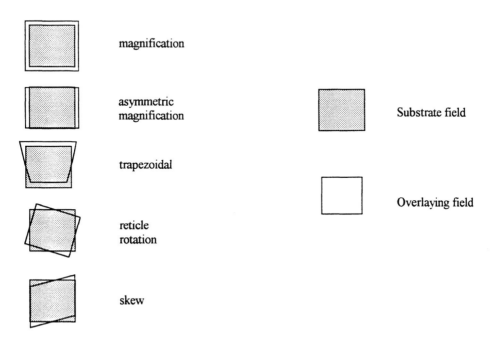

Figure 5.3. Intrafield overlay errors.

Intrafield overlay models for wafer steppers were introduced by MacMillan and Ryden in 1982. Since then, the most significant change has been the introduction of models for step-and-scan systems. The models for step-and-scan systems include parameters for asymmetric magnification and skew, which are not relevant for step-and-repeat systems. Suppose the same stepper is used for printing two layers, one overlaying the other. Let (x, y) be the coordinates of a point on the wafer, relative to the center of the exposure field in which it is contained. For a step-and-repeat system the intrafield overlay errors are modeled:

$$\delta x = mx - ry + T_x xy + T_y x^2 + e_x \tag{5.8}$$

$$\delta y = my + rx + T_y xy + T_x y^2 + e_y \tag{5.9}$$

In these equations, the parameter m represents a magnification error. It is dimensionless and is usually expressed in parts-per-million. The magnification error can be interpreted as follows. The error in overlay increases linearly with distance from the center of the exposure field. The rate at which this error increases with distance from the center of the exposure field is the magnification error coefficient m. For step-and-repeat systems the magnification error is the same in x and y directions. The coefficient r represents reticle rotation. As with wafer rotation, this is usually expressed in microradians. Reticle rotation error also increases linearly with the distance from the center of the exposure field, and the error coefficient is the same for errors in x and y. The factors T_x and T_y are trapezoidal errors. These types of overlay errors occur on lenses that are not telecentric on the reticle side. The terms e_x and e_y are residual errors that represent the intrafield overlay that does not conform to the model.

Interfield models are identical for step-and-repeat and step-and-scan systems, but the appropriate intrafield models differ. For step-and-scan systems, the intrafield model is:

$$\delta x = m_x x - r_x y + e_x \qquad (5.10)$$

$$\delta y = m_y y + r_y x + e_y \qquad (5.11)$$

The step-and-scan model is similar to the step-and-repeat model for doubly telecentric lenses, with some differences. Most notably, the magnification coefficents (m_x and m_y) and rotation coefficients (r_x and r_y) in Eqs. 5.10 and 5.11 are independent in the two equations, and not coupled, as in Eqs. 5.8 and 5.9. The magnification for step-and-scan systems can be different in the x and y directions, because the magnification in the x direction has a different origin than the magnification in the y direction. The magnification in the direction of the scan is determined by the relative speeds of the reticle and wafer stages. Ideally, the reticle stage should move N times faster than the stage, where $N{:}1$ is the reduction ratio of the lens (typically, $N = 4$). When the reticle stage speed moves faster than N times the wafer stage speed, the printed field will be shorter in the scan direction. The magnification in the direction perpendicular to the scan is determined only by the lens reduction factor ($N{:}1$). For step-and-scan systems, pure reticle rotation occurs when the scans of the reticle and wafer stages are parallel, but the reticle is rotated relative to both. Field skew results when the reticle is rotated properly with respect to the wafer, but the reticle stage is rotated.

For any point on the wafer the overlay error is the sum of the interfield and intrafield errors. In the X direction, the total overlay error is $O_{X,x} = \Delta X + \delta x$, and the overlay error in the Y direction is $O_{Y,y} = \Delta Y + \delta y$, at the point $(X + x, Y + y)$. For step-and-scan systems:

$$O_{X,x} = T_X + E_X X - R_X Y + m_x x - r_x y + \rho_{X,x} \qquad (5.12)$$

$$O_{Y,y} = T_Y + E_Y Y + R_Y X + m_y y + r_y x + \rho_{Y,y} \tag{5.13}$$

where $\rho_{X,x}$ and $\rho_{Y,y}$ are total residual errors. There are similar equations for step-and-repeat systems.

Classifying errors is important for controlling overlay, because intrafield and interfield overlay errors generally arise from different causes. For example, reticle rotation errors involve the alignment of the reticle, while wafer rotation issues involve water alignments. In order to classify overlay errors the coefficients in Eqs. 5.12 and 5.13 need to be determined. The most common method for extracting these coefficients from measured data is the method of least-squares.[6,7] In this technique, more data are needed than there are parameters. For Eqs. 5.12 and 5.13, measurements need to be made at two points per exposure field, or more. (Each measurement site provides two measurements, one in X and one in Y.) The least squares method determines the set of coefficients $\{T_X, T_Y, m_x, ...\}$ which minimizes the sum of squares of the residual errors:

$$\sum_{X,x,Y,y} \left(\rho_{X,x}^2 + \rho_{Y,y}^2 \right), \tag{5.14}$$

where the sum is extended over all measured points. Computational speed and insensitivity to noise are among the reasons why the least-squares method is the method most commonly used. The choice of mathematical methods used to determine the model coefficients will be discussed further later in this chapter.

In Section 3.3 it was noted that the trending of process parameters was useful in circumstances in which test wafers were required. For overlay, these process parameters are the coefficients in the overlay model. By examining the variation in these coefficients, the sources of variation can often be identified. For example, large excursions might be correlated to the stepper used for a particular prior layer.

Example: An engineer wanted to compensate for reticle registration errors that had been measured on a product reticle. Errors on the reticle that fit the model of Eqs. 5.10 and 5.11 could be corrected on step-and-scan systems. Constants were added to each of these equations as well, to correct for translation shifts. The reticle errors, as measured on the product reticle, are given in Table 5.1. The model coefficients are given in Table 5.2. As can be seen, a substantial fraction of the reticle error was eliminated by applying magnification, rotation, and translation offsets. For example, the registration error at the position (-69.5, 0) is given by:

$$\delta x = X_{\text{translation}} + m_x x - r_x y + e_x \tag{5.15}$$

$$= -10.6 + 0.963 \times 69.5 - 0.329 \times 0.0 + e_x \tag{5.16}$$

$$= 56.3 + e_x \qquad\qquad (5.17)$$

The measured error was 63 nm, so the model was able to account for most of the registration error at that point, and most of the error was therefore correctable.

Measurement positions		Measured registration errors		Residual registration errors Step-and-scan model	
x (mm)	y (mm)	δx (nm)	δy (nm)	e_x (nm)	e_y (nm)
-69.5	0.0	63.0	0.0	6.7	-9.7
69.5	0.0	-63.0	0.0	14.5	-12.9
-30.2	51.0	17.0	-15.0	15.3	-2.2
30.2	51.0	-14.0	1.0	42.4	12.4
30.2	-51.0	-21.0	30.0	1.9	-5.4
-30.2	-51.0	48.0	29.0	12.8	-5.0
-29.9	50.8	-41.0	-16.0	-42.5	-3.3
29.9	50.8	-82.0	-7.0	-25.9	4.3
29.9	-50.8	-61.0	39.0	-38.3	3.7
-29.9	-50.8	48.0	52.0	13.1	18.1

Table 5.1. Reticle registration errors.

The first two columns are the positions where registration was measured. The third and fourth columns are the registrations errors, as measured, and the last columns are the residual errors, after the modeled part of the errors was subtracted from the raw data.

x-translation	-10.6 nm	y-translation	11.3 nm
m_x	-0.963 ppm	m_y	0.023 ppm
r_x	0.329 µrads	r_y	-0.459 µrads

Table 5.2. The model coefficients, determined by least squares, for the reticle mis-registration data in Table 5.1.

5.2 MATCHING

A set of non-random overlay errors that are unaccounted for in the overlay models are found in situations where more than one stepper is used. These errors are referred to as matching errors, and there can be grid and intrafield matching errors. Grid matching errors arise from absolute stepping errors. While stepper stages are extremely repeatable, their stepping may deviate at each stage position, on average, from a perfect grid. When a single stepper is used for both the substrate and overlaying layers, these average deviations cancel out. A different situation arises when different steppers are used for the two layers.

Grid registration errors originate in the non-flatness of the stage mirrors. Consider the stage mirror drawn in Fig. 5.4. The branch of the interferometer that measures the stage's position in the Y-direction will measure apparent movement in the Y-direction as the stage is moved from left to right, because of the non-flatness of the mirror. When a single stepper is used, the resulting stepping errors, relative to a perfect grid, usually do not lead to overlay errors, because these stage stepping errors occur for all layers and cancel out. In general, the mirror non-flatness varies from stepper to stepper, and stage mismatch occurs. For modern wafer steppers the stage mirrors are flat to approximately $\lambda/40$, where $\lambda = 632.8$ nm is the wavelength of light of the stage's interferometer, or about 16 nm. In addition to not being flat, the mirrors used for the interferometry in the X and Y directions will not be exactly orthogonal, with errors on the order of one arcsecond.

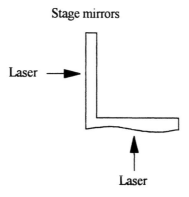

Figure 5.4. Non-flat stage mirrors, which lead to matching errors.

Grid registration errors can be corrected using software. Typically, reference matching wafers are used for matching all steppers in a facility to the same positions. Look-up tables can be used for these corrections, or the grid models can be expanded to higher orders. The second order interfield term is wafer bow.[8] Matching grids to less than 5 nm is not uncommon. Laser interferometers for wafer steppers usually have more than one beam for each direction, in order to measure stage rotation during stepping. Mirror non-flatness can then lead to intrafield rotation as fields are stepped across wafers,[9] which will not be captured by models which treat the reticle rotation the same in every field.

Lenses do not always place geometries where they belong. For example, suppose the pattern on the reticle consisted of a perfect rectangular grid (Fig. 5.5). Because of the lens placement errors the vertices of the pattern will be slightly moved. Placement errors run from 200 nm on older lenses to < 20 nm on contemporary machines. These errors may be inherent in the design or result from fabrication imperfections. The design contribution is referred to as lens distortion, and has a known functional form:

$$\Delta r = D_3 r^3 + D_5 r^5 \tag{5.18}$$

$$\Delta x = D_3 x r^3 + D_5 x r^4 \tag{5.19}$$

$$\Delta y = D_3 y r^2 + D_5 y r^4 \tag{5.20}$$

where r is the radial distance from the center of the lens field:

$$r = \sqrt{x^2 + y^2} \ . \tag{5.21}$$

Because of the magnitude of the exponents of terms in Eq. 5.18, D_3 and D_5 are referred to as third and fifth order distortion, respectively.

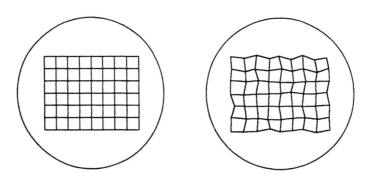

Perfect pattern on the reticle.

Pattern on the wafer, distorted as a consequence of lens placement errors

Figure 5.5. Pattern distortion as a consequence of lens placement errors. The movement of the vertices is exaggerated in this figure for the purpose of illustration.

Distortion tends to be fairly constant within a particular lens family. However, there are strong economic reasons to minimize the numbers of very high performance steppers, and to mix-and-match steppers of different types.[10]

Consequently, lenses of more than one type will usually be used to fabricate semiconductor devices. Older generations of projection optics had non-zero levels of distortion in the design, where the magnitude of Δx and Δy due to distortion alone could be as large as several tens of nanometers, up to 200 nm. In such situations, good matching could be achieved only within given lens families, particularly when different lens designs might have distortions with opposite signs for D_3 and D_5. The designs of modern lenses have negligible levels of radial third and fifth order distortion. For these newer lenses, placement errors are due primarily to manufacturing imperfections, which produce intrafield errors that are not systematic, as in the case of third and fifth order lens distortion. Polishing variations across the surfaces of lenses and mirrors and inhomogeneity in optical materials can lead to intrafield registration errors, and these will tend to have random patterns.

In the presence of significant radial lens distortion it is impossible to distinguish between third and fifth order distortion and simple magnification errors, unless a sufficient number of points are sampled within the exposure field. In-line sampling plans usually consist of measurements at the four corners of the exposure field, perhaps with measurements also near the center of the field, and this is insufficient for distinguishing between magnification and higher order radial distortion. The problem of small samples for in-line process control will be considered shortly in more detail.

While distortion is extremely stable for i-line systems, distortion varies with very small shifts — less than a picometer — in wavelength for DUV systems.[11, 12] Consequently, additional controls are needed on DUV machines, the most critical being wavelength control. Modern KrF excimer lasers, the light source for 248 nm exposure systems, contain internal wavelength references that provide good wavelength control.[13] As with anything, this calibration system can malfunction or drift, and there is always the potential for changes in third and fifth order distortion as a consequence of a small shift in wavelength. Also, changes in barometric pressure will change the wavelength of light in the air, and excimer steppers must adjust for this correctly, or third and fifth order distortion may be introduced.

For step-and-scan systems the intrafield placement errors will also have a scanning component. There will be random and systematic contributions to the scanning errors. This random component does not exist in step-and-repeat systems, so there is more potential for overlay errors with scanners, even for a stepper-to-itself. Moreover, the stage which scans the reticle can malfunction, another control problem unique to step-and-scan systems.

The intrafield overlay will contain some theoretically correctable contributions which can be fit to the overlay model (Eqs. 5.8 - 5.11), and other components that arise from differences between lenses which do not conform to the model. The matching of the lenses is the set of residual overlay vectors which remains after subtracting out the correctable contributions. These sets of vectors are dependent upon the criterion used for determining the model parameters. The least squares method is the one used most commonly, but it is not the only criterion that is reasonable to use. For example, since devices are usually designed to yield so long as overlay is less than a particular value and fail when this value is exceeded, yield can be maximized when the worst overlay errors are minimized.[14] This criterion consists of the minimization of:

$$\max_{x,y}\left(\left|e_x\right|,\left|e_y\right|\right),\qquad\qquad(5.22)$$

where e_x and e_y are the residual errors in Eqs. 5.8 – 5.11. This differs from the least-squares criterion, which consists of the minimization of the quantity in Eq. 5.14 and will typically lead to different model coefficients. We say that lenses are matched when correctable parameters have been set to minimize the selected criterion, either Eq. 5.14, 5.22, or another that may have been chosen.

The least-squares method has two key advantages. The first is computational simplicity, since model coefficients can be found through simple matrix calculations. The least-squares method also provides transitivity,[15] that is, if Stepper A is matched to Stepper B, and Stepper B is matched to Stepper C, then Stepper A will be matched to Stepper C. Minimization of maximum error does not share this characteristic.[16]

When only a single stepper is used, the intrafield overlay model accounts for nearly all of the intrafield overlay errors, and the overlay errors throughout the entire field can therefore be inferred from measurements at only a few points with the exposure fields. It is quite a different situation when more than one stepper is used. Matching methodology must account for overlay in the areas of the exposure field that are not accessible when measuring product wafers. Specialized structures are normally used for measuring overlay, and these are usually placed in the scribe lanes between product dies. This limits the amount of overlay data collected on product wafers, since most of the exposure field is occupied by product devices, not overlay measurement structures. This limitation on the number of points within each exposure field at which overlay can be sampled has significant consequences, since overlay is measured in areas where the product is not placed.

Suppose overlay is measured on several fields on the wafer, and within each field at four sites with the same intrafield positions. One can fit the acquired data to the overlay model (Eqs. 5.12 and 5.13) by minimizing the appropriate metric (least squares, minimum error, etc.). This approach may not optimize the overlay over the entire exposure field, particularly the parts of the field in which the product is located and overlay is not measured. This was demonstrated in the following way.[17] Between two steppers the overlay was measured on a 12 × 11 grid with 1.95 mm spacings in the x and y directions within the exposure field. The resulting lens matching is shown in Fig. 5.6. In a *gedanken* (thought) experiment the overlay measurements were considered at only four points among these 132 sites, in a pattern that resembled a typical overlay measurement plan. Several subsets of four points were considered

(Fig. 5.7). For each set of four sites the overlay model coefficients were re-calculated and plotted in Fig. 5.8. As one can see, the resulting model coefficients varied significantly among sampling plans. Recall that the baseline set of coefficients was the one that optimized the overlay over the entire exposure field, not just at the four corner points. By measuring at only four points and adjusting the stepper to minimize overlay at just the measured points, overlay is moved away from the optimum, overall. This is a problem that occurs when different steppers are used for the overlaying patterns.

In earlier discussions it was mentioned that the production of good devices requires that critical parameters remain within specifications at all locations within the dies. Lens mismatch may result in particular points at which overlay is very poor. Overall overlay may be very good, but the die at the field location where the mismatch occurs will yield poorly. An example of this is shown in Fig. 5.6, where the upper left corner of the field has a point with particularly bad overlay. Overall, the overlay is good, but the die located in that corner will consistently have poor overlay. With $1 - 8$ die per field being typical, a single bad site within the exposure field can degrade the yield significantly. The concept that yield requires all points within a die to be within specifications was first introduced in the context of overlay.[14]

To complicate matters further, it has been observed that lens placement errors vary with illumination conditions, such as partial coherence[18,19,20] or off-axis illumination.[21, 22] For purposes of maximizing depth of focus or exposure latitude, different illumination conditions are often used at different layers. For example, large values for the partial coherence are used for gate and metal layers, at least when binary masks are used. Small values of partial coherence are used for contact layers and phase shifting masks. Consequently, there will be intrafield errors, even when the same stepper is used for all layers, if different illumination conditions are used for different layers. The overlay errors between a first layer printed with standard illumination (NA = 0.63, σ = 0.65) and a second layer exposed using the same numerical aperture, but a partial coherence = σ = 0.3, are shown in Fig. 5.9.[20] Even though both layers were printed using the same lens, there is considerable intrafield error. Each vector is the overlay error at the point located at the tail of the vector. It should also be noted that the use of a single stepper does not guarantee the non-existence of grid and intrafield matching errors, because drift, database corruption, and stepper malfunction are possible. However, such events are rare, and the matching errors discussed here occur infrequently when single steppers are used with fixed operating parameters.

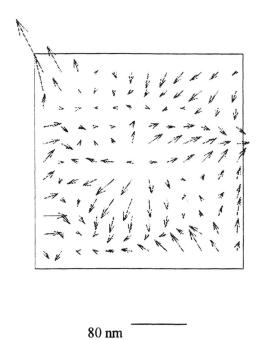

80 nm

Figure 5.6. The lens matching between a step-and-scan system and a step-and-repeat system. The vectors are the residual errors after subtracting correctable errors, at the points located at the tails of each vector.

1	3	5	8	10		13	9	8	5	3	1
2	6	12	15	18			18	15	12	6	2
4	11	17	20	22			22	20	17	11	4
7	14	19	23	25			25	23	19	14	7
9	16	21	24	26			26	24	21	16	10
13											13
10	16	21	24	26			26	24	21	16	9
7	14	19	23	25			25	23	19	14	7
4	11	17	20	22			22	20	17	11	4
2	6	12	15	18			18	15	12	6	2
1	3	5	8	9	13		10	8	5	3	1

Figure 5.7. Sampling of the measurement site of the data shown in Fig. 5.6. For example, the four corner sites were chosen for Sampling Plan 1.

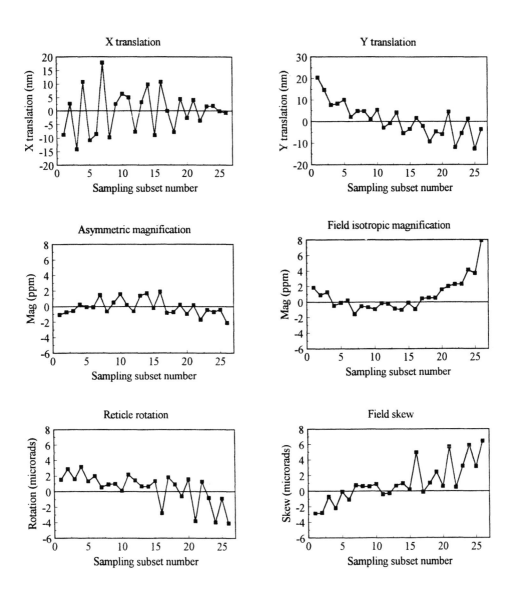

Figure 5.8. Variations in model parameters for the sampling plans shown in Fig. 5.7.

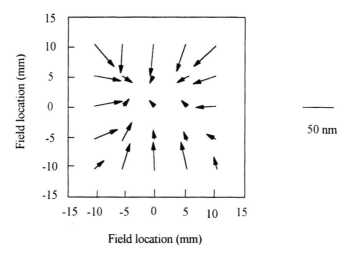

Figure 5.9. Intrafield overlay errors, where the first level was printed on a stepper with NA=0.63 and σ = 0.65, and the second level was printed with the same NA and σ = 0.3.

The changes in lens placement errors can be understood by considering the imaging of a simple diffraction grating, discussed in Chapter 4. Depending on the angle of incidence of the illumination, numerical aperture and the pitch of the pattern, the light of a grating pattern projected through a lens will pass through different parts of the lens. Many aberrations are a result of variations in the polishing of optical surfaces and inhomogeneity in glass materials. Light rays going through one part of the lens will have different errors than rays going through different parts of the lens. As a consequence, aberrations will vary for light passing through different parts of the lens. For a given feature, the aberrations of its image will depend upon the particular imperfections in those parts of the lens through which the light from that feature passes. Hence, aberrations will vary with pitch, numerical aperture, and the illumination conditions. The light from patterns other than gratings will also be distributed throughout the optics with dependence upon feature size and proximity to other features.

As feature sizes shrink, these subtle issues will generate overlay errors of significance. For example, certain aberrations, such as coma, result in a feature-size dependency for intrafield registration. This is shown in Fig. 5.10, where simulated image placement shifts of isolated clear lines are plotted versus spacewidth and where the lens has $a_7 = 0.035$ waves of coma.[23] Overlay measurement structures with large features will measure different overlay than actually occurs for critical fine-linewidth features in the product, in the presence of coma. This implies that overlay measurements will not represent the overlay of critical features in the circuits. Measuring overlay using small features directly is a problem. Overlay is typically measured using optical tools (Chapter 8) that can measure features reliably only if their size is greater than about one micron. However, optical tools have throughput and cost advantages over measurement equipment capable of measuring smaller features. Subtle issues such as these will need to be addressed when 10 nm becomes a significant fraction of the total overlay budget.

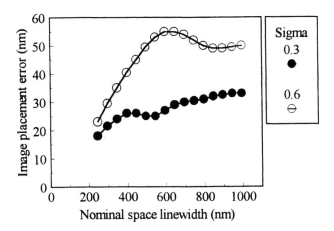

Figure 5.10. Simulated image shifts for 0.5 NA optics at a wavelength of 248 nm, and $a_7 = 0.035$ waves of coma.

In order to enhance productivity, high throughput steppers with large exposure fields are often used for non-critical layers, while smaller field steppers are used for the critical layers. The mixing of these different types of steppers often results in non-concentric exposure fields (Fig. 5.11). Control of overlay requires extensions of the models discussed thus far, that assume that the centers of exposure fields sit approximately on top of each other.[24,25] Overlay needs to be measured at the corners of the critical fields, in order to identify overlay errors associated with intrafield errors associated with the smaller field. Mathematical models describing field rotation and magnification must take into account the non-concentric nature of the fields.

Misregistration of geometries on reticles will result in overlay errors on wafers. This misregistration can have random components, as well as other contributions which vary systematically across the reticles. An example of a systematic variation is one which varies linearly across the reticle. Such linearly varying misregistration will appear as a magnification error. It is possible to correct for systematically varying reticle errors which correspond to adjustable intrafield parameters.[122, 26] For example, reticle misregistration which results from orthogonality errors of the reticle beam writer will correspond to field skew. Such an error is correctable on a step-and-scan system, but not a step-and-repeat machine. Correction of reticle errors was described in an example shown in the prior section.

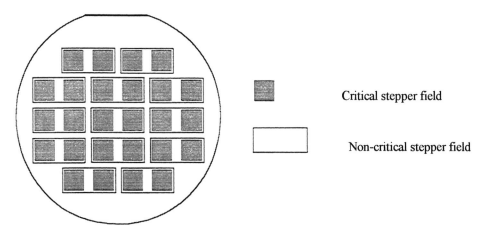

Figure 5.11. Wafer layout when a wide field stepper is mixed and matched with a smaller field critical layer stepper.

There remain other overlay errors that are not accounted for in the overlay model. Wafers can experience non-linear plastic distortions, often caused by rapid thermal processing,[27] which will not be corrected by alignment methods that account only for linear errors. Wafer flexure, which could be caused by a particle under the wafer, will also cause overlay errors which do not conform to the overlay models discussed, but which are nevertheless quite real. Consider the situation depicted in Fig. 5.12. Flexure of the wafer will result in lateral pattern displacement, according to the following equation[28]:

$$\Delta L = \frac{t\theta}{2}, \tag{5.23}$$

where t is the thickness of the wafer. We can estimate the magnitude of this overlay error from elasticity theory. A circular plate clamped at the edges and displaced by an amount Δ_{center} at its center will be displaced vertically at radius r, measured from the center of the circle, by the amount[29]:

$$\Delta(r) = \Delta_{center} \frac{2}{a^2}\left(r^2 \ln\frac{r}{a} + \frac{1}{2}\left(a^2 - r^2\right)\right), \tag{5.24}$$

where a is the radius of the circle. The maximum slope for this displacement is

$$\theta_{max} = \frac{4\Delta_{center}}{ae}, \tag{5.25}$$

where e is the base for natural logarithms. Suppose a particle on the chuck causes a maximum vertical displacement of 0.2 μm, affecting a circular area of radius $a = 5$ mm. Then $\theta_{max} = 5.89 \times 10^{-5}$ radians. For SEMI standard 200 mm wafers, $t = 750$ μm. From Eq. 5.23, the resulting overlay error is 22 nm. For 180 nm technology, where 65 nm overlay is required, this is a large fraction of the overlay budget.

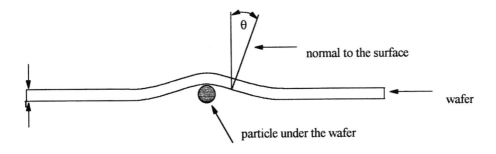

Figure 5.12. Wafer deformed because of a particle between the wafer and the chuck of the stepper.

5.3 CONTRIBUTIONS FROM PROCESSING AND ALIGNMENT MARK OPTIMIZATION

The quality of the overlay is dependent upon the ability of the stepper's alignment system to acquire alignment targets accurately. There is an interplay between the alignment signal and the nature of the alignment targets, which depends upon the overall process for making semiconductors, including film depositions, resist coatings, etches, and polishes. Consequently, lithography engineers need to optimize alignment targets. This optimization will include the dimensions of targets, etch depths, resist coatings, and polishing processes.

The spin coating of resist over topography does not produce symmetrical films,[30, 31] as illustrated in Fig. 5.13. Optical alignment systems will be susceptible to thin film interference effects which result in low-contrast signals[32] or misalignment of non-planar alignment marks.[33,34] Because spin coating is a radial process, the misalignment will go in opposite directions on diametrical sides of the wafer and result in a scaling error. Resist coating processes can be modified to reduce the impact of thin film optical effects.[30, 35] Optical modeling has shown that dark field alignment systems are less sensitive to these thin film optical effects than bright field or diffraction alignment systems.[36] On the other hand, the graininess of highly reflective metal, polysilicon or silicide surfaces introduces considerable noise into dark field alignment systems. For metal surfaces, bright field alignment systems are preferable. The influence of grains can be minimized by the use of low coherence illumination (large σ) in the alignment optic or interference techniques, which involves the design of the alignment system and is generally beyond the control of the user.[37]

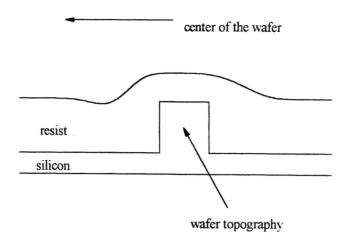

center of the wafer

resist

silicon

wafer topography

Figure 5.13. Resist asymmetry as a consequence of resist spin coating.

Sputter deposition of metals can cause asymmetries in overlay targets. Metal ions will be partially shadowed by alignment mark topography, leading to apparent shifts in the position of the alignment mark (Fig. 5.14). Because sputtering geometries are usually radially symmetric, the resulting overlay errors will appear as wafer scaling errors. The overlay measurement structures may be affected in the same way as the alignment targets, and the overlay errors may be apparent only after etching the metal. The asymmetry is affected by the geometries of substrate features, and it may be different for large structures used to measure alignment and critical features in the product. If the asymmetries are consistent within a lot, and lot-to-lot, then overlay offsets may be determined off-line and used for correction.

Chemical-mechanical polish (CMP) represents a particular challenge for overlay.[38, 39] The purpose of CMP is to produce highly planarized surfaces. While this helps lithography by reducing thin film effects and increasing depth of focus, it makes overlay difficult because it reduces alignment target contrast. Consider the situation depicted in Fig. 5.15a. Alignment targets are invisible when planarization is complete and wafers are covered completely by metal films. The polishing process can be adjusted in order to produce some degree of topography (Fig. 5.15b),[40] but the polishing often produces asymmetries (Fig. 5.15c), particularly when the width of the pattern is large. Alignment target deformation can occur from all CMP processes, such as for shallow trench isolation, not just those involving metal layers.

There are several things that can be done to improve overlay in the context of CMP. Alignment mark dimensions can be varied or segmented in order to determine the best mark. Polishing engineers should participate in any programs to improve overlay, since modifications of the polish process can often reduce overlay errors considerably, particularly when such errors result from highly variable polish processes. For example, the use of harder polishing pads may lead to reduced overlay errors. Stepper manufacturers have also responded with modifications to their alignment systems, using modified illumination[41] and new algorithms for interpreting alignment signals.[42] It is also important to work cooperatively with one's stepper supplier, since alignment system redesigns can result as the stepper manufacturer learns more about the actual use of the system.

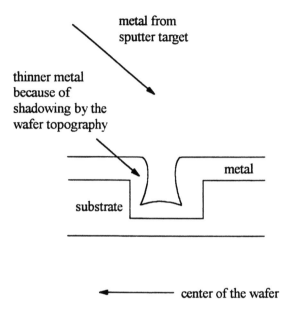

Figure 5.14. Asymmetry as a consequence of the geometry of metal deposition.

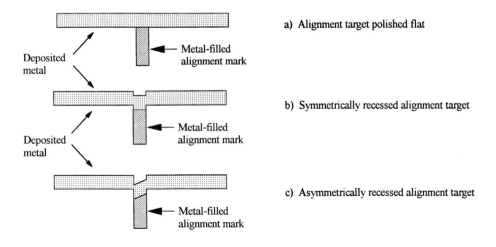

Figure 5.15. Alignment targets following chemical-mechanical polish.

5.4 ADDRESSING THE PROBLEM OF NON-NORMAL DISTRIBUTIONS

Distributions of overlay measurements from individual lots are typically very non-normal. Characterization of overlay and the dispositioning of lots must take this into account. Consider, for example, the overlay distribution shown in Fig. 5.16. The overlay had been measured at the four corners of five exposure fields. In this situation the overlay specification was 120 nm. It can be seen that all of the measurements are well contained within these specification limits. Nevertheless, the +/- 3σ points are approximately 150 nm, significantly larger than the specification limit. Clearly the 3σ points cannot be compared to the specification limits in order to disposition lots, and it is therefore not obvious how process capability can be assessed. The implementation of statistical process control also becomes complicated because of non-normality, as noted earlier. The issues of lot dispositioning, SPC, and process capability in the context of overlay will be discussed further in this section.

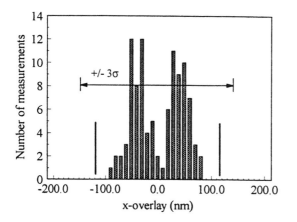 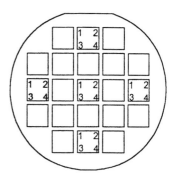

Figure 5.16. Overlay distribution from a lot of wafers with a 3 ppm field magnification error. Vertical lines are located at the specification limits of +/- 120 nm. The sampling plan is shown on the right.

When measurement distributions are normal, lot dispositions and the implementation of statistical process control is straightforward. This is because we know the functional form of the distribution, and it is completely characterized by only two parameters, the mean and the standard deviation. With overlay we also know the functional form, given by the overlay models discussed in the previous section. It is, however, much more complicated than a normal distribution (Table 5.3).

Normal distribution	Step-and-repeat overlay model	Step-and scan overlay model
Mean	X- & Y-translation	X- & Y-translation
Standard deviation	X- & Y-scale	X- & Y-scale
	Wafer rotation	Wafer rotation
	Orthogonality	Orthogonality
	Magnification	Magnification
	x- & y- trapezoid	Asymmetric magnification
	Reticle rotation	Skew
		Reticle rotation

Table 5.3. Parameters of different distributions.

Lot dispositioning can be accomplished in several ways. The simplest method is based upon the largest overlay error, which is either the maximum overlay error or the absolute value of the minimum error, whichever is biggest. This approach is simple and works reasonably well. Clearly, if there are measurements that exceed specification limits, then reworking the lot should be considered. This method does have some drawbacks that should be noted. First, the sampling must cover enough of the wafer. If there are wafer scaling errors, then the worst overlay will be on the outermost portions of the wafer. However, measurements near the edges of wafers are often unreliable, as a consequence of non-uniformities in etch, film deposition, and polish. Bad measurements, among other problems, can lead to outliers in the data, to which criteria based upon extreme values are very sensitive. Methods for handling outliers will be discussed later in this chapter.

A more sophisticated method for lot dispositioning directly involves the parameters of the overlay model. It is good practice to calculate and track the overlay model parameters, and this is easily accomplished with readily available measurement equipment and overlay analysis software. Limits can be placed on all of the individual parameters, and lots will be considered to have failed to meet specifications if any parameter exceeds its limits.

Appropriate limits for the individual model parameters can be determined through overlay simulations. Let us assume that each parameter is normally distributed within each lot (wafer-to-wafer), and that the mean parameter value varies normally lot-to-lot. One can then run a Monte-Carlo simulation to see what the resulting overlay distribution might be.[43] The results of one such simulation are shown in Fig. 5.17, using variations in the model parameters shown in Table 5.4. This simulation was performed assuming that each parameter in the overlay model (Eqs. 5.12 and 5.13) varied according to a normal distribution. The "lot-to-lot" variation provided the mean error for the wafers within each "lot," for each parameter. The stage precision listed in Table 5.4 gave the residual errors for Eqs. 5.12 and 5.13. For example, the average translation error for the first "lot" was selected randomly, assuming that the translation error varied normally, with a mean of zero and a 3σ variation of

30.0 nm. The "wafers" in this "lot" had this randomly selected translation error, plus additional translation error which varied "wafer-to-wafer" (15.0 nm, 3σ). Every parameter in the overlay model was determined similarly. The simulation was repeated over thousands of "lots" and "wafers."

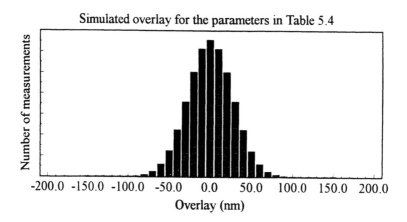

Figure 5.17. Simulated overlay, using the parameters of Table 5.4. The "sampling" plan is the same one shown in Fig. 5.16, with the addition of measurements in the centers of the sampled exposure fields.

In this simulation, the overlay was "measured" at five points per 20 mm × 20 mm field (the corners and center), on five fields on the 200 mm diameter wafers. This distribution of measurement sites determined the values for X, Y, x and y in Eqs. 5.12 and 5.13. The results of this simulation are shown in Fig. 5.17. The resulting overlay has only a few ppm measurements that exceed 120 nm, so the values in Table 5.4 could be used as specification limits for the overlay parameters, for a process with 120 nm overlay requirements. There are numerous combinations of specification limits on the individual overlay parameters that are consistent with a total overlay budget. Assigning limits to each parameter is an interative process, where equipment capability is used to balance parameters in order to minimize rework.

To generate Pareto charts for overlay, contributions need to be converted to common units. Even though translation errors typically have units of nanometers, rotations are measured in micro-radians, and scale factors are in terms of parts-per-million. There are several ways to compare them. The simplest method is to convert the rotation and scale factors to equivalent nanometers. A reticle rotation error of 1.2 μrad across a 20 mm square field contributes up to ± 12 nm overlay error in x and y. A more sophisticated approach involves the use of overlay models. An overlay simulator can be run to replicate an existing process. One could then change each parameter in Table 5.2, one at a time, to zero. The contribution of each parameter to overlay is the difference between the total overlay obtained with the zero and non-zero parameters.

Parameter	Wafer-to-wafer (3σ)	Lot-to-lot (3σ)	
Grid scale	0.1	0.3	ppm
Orthogonality	0.1	0.3	Microradians
Wafer rotation	0.1	0.3	Microradians
Translation	15.0	30.0	Nanometers
Field magnification	3.0	3.0	ppm
X-magnification	3.0	3.0	ppm
Reticle rotation	3.0	3.0	Microradians
Field skew	3.0	3.0	Microradians
Stage precision	15.0	0	nm

Table 5.4. Parameters for the simulated results shown in Fig. 5.17.

If overlay parameters are going to be used for critical decisions, then the overlay needs to be measured in ways that minimize the uncertainties in the model parameters. How this is accomplished can be understood from consideration of a simple example. Suppose one is fitting to a linear model involving a single independent variable X:

$$Y = AX + B, \tag{5.26}$$

using data collected in pairs:

$$\left(X_1, Y_1\right), \left(X_2, Y_2\right), \ldots, \left(X_n, Y_n\right) \tag{5.27}$$

Suppose that the Y_i are independent, and that the residual errors

$$\delta Y_i = Y_i - AX_i - B \tag{5.28}$$

vary with the standard deviation σ. If the fit of the model is optimized by the criterion of least squares, then[44]:

$$\text{Standard deviation of } A = \sqrt{\dfrac{\sigma^2}{\displaystyle\sum_{i=1}^{n}\left(X_i - \overline{X}\right)^2}} \tag{5.29}$$

$$\text{Standard deviation of } B = \sqrt{\frac{\sigma^2 \sum_{i=1}^{n} X_i^2}{n \sum_{i=1}^{n} (X_i - \overline{X})^2}} \quad \xrightarrow{\overline{X}=0} \quad \frac{\sigma}{\sqrt{n}} \quad (5.30)$$

From these expressions, two rules for obtaining good model parameter estimates can be seen:

1) The best estimates are obtained by maximizing the separations among the sampling points X_i.

2) The uncertainty in the estimates of the model parameters decreases as the square root of the number of measurements.

These results are similar for situations involving more than one independent variable, and have been observed on measurements on products.[45]

In order to calculate all linear grid parameters, one needs to measure overlay in at least three non-collinear fields. This provides six pieces of data (X and Y measurements), sufficient for computing the six grid parameters, X and Y translation, X and Y scaling, wafer rotation, and orthogonality. By measuring equivalent field locations one can avoid confounding interfield and intrafield contributions. However, measurements in only three fields are usually inadequate, because of noise. Eq. 5.30 is a useful guide for determining the minimum number of measurement fields needed. In order to calculate scaling, rotation, and orthogonality parameters it is useful to locate the measurement sites close to the edge of the wafer, while avoiding anomalies that may occur very close to the wafer edge.

The calculation of field terms also requires a minimum of three sites. As with the grid parameters, the accurate calculation of the model coefficients requires that the measurement sites be located as close to the outside edge of the exposure field as possible. As discussed earlier, the in-line measurements are limited to locations in scribe lanes between product dies, and the results need to be interpreted accordingly.

Multivariate control charts provide an alternative to charting every model parameter,[46] which can provide a substantial reduction in the number of control charts required and the number of false alarms. However, since an out-of-control condition will require that the specific out-of-control parameters be identified, it will still be necessary to maintain databases of the individual overlay parameters.

5.5 OUTLIERS

The problem of outliers, where one or several data points do not appear to belong to the same distribution as the other data, is illustrated in Fig. 5.18. There are data which clearly lie outside of the distribution of the rest of the data. This can result from a number of sources. Metrology can often be the cause of outliers. In this case the anomalous data can be disregarded in terms of assessing the true quality of the process. Defects can also be the cause of measurements which exceed control or specification limits. Particles on the backsides of wafers can cause high values for overlay, as discussed earlier. If such particles are attached to the backsides of individual wafers, then the number of lost dies due to the particles may be small, unless these particles occur repeatedly. If the particle is attached to the chuck of the stepper, then the poor overlay will be occurring on all wafers processed on that tool, significantly degrading yield. In this latter situation, there would be a significant penalty for ignoring the outliers.

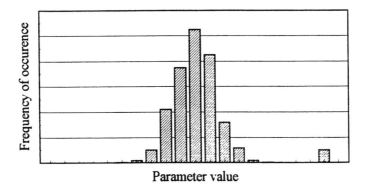

Figure 5.18. Histogram of data containing outliers.

Clearly, outliers must be handled carefully. The first step toward dealing with outliers is identifying them as such. Simply because data exceed control limits does not mean that there are outliers, since the process may well have gone out of control. Tests that identify the presence of outliers are referred to as discordancy tests.[47] A number of tests have been devised to detect outliers in common distributions. For example, the following statistic can be used to identify the presence of an outlier in an otherwise normal distribution:

$$T = \max\left(\frac{x_{max} - \bar{x}}{s}, \frac{\bar{x} - x_{min}}{s}\right),\qquad(5.31)$$

where s is the sample standard deviation. If T exceeds the values in Table 5.5, then one can assume that there is an outlier present.

Eq. 5.31 and Table 5.5 provide the information necessary to identify an outlier in an otherwise normal distribution. As discussed earlier, the data obtained in overlay measurements often form non-normal distributions. For overlay, outliers can be identified by analyzing the residual errors. That is the data which remain after subtracting out the modeled components:

$$\rho_{X,x} = O_{X,x} - \left(T_X + E_X X - R_X Y + m_x x - r_x y \right), \tag{5.32}$$

where $O_{X,x}$ is the measured overlay error in the X direction. Since the residuals $\rho_{X,x}$ tend to be normal, discordancy tests such as Eq. 5.31 can be applied. Many overlay analysis programs automatically reject outlier data when computing model parameters. As mentioned previously, fully automatic rejection of outliers can prevent the identification of real and persistent problems, such as particles on stepper chucks.

n	95% confidence	99% confidence
3	1.15	1.15
4	1.48	1.50
5	1.71	1.76
6	1.89	1.97
7	2.02	2.14
8	2.13	2.28
9	2.21	2.38
10	2.29	2.48
12	2.41	2.63
15	2.55	2.81
20	2.71	3.00

Table 5.5. Values for the parameter T, for the purpose of identifying outliers.

---

[1] Semiconductor Equipment and Materials International, Mountain View, California.

[2] N. Magome and H. Kawai, "Total Overlay Analysis for Designing Future Aligner," SPIE Vol. 2440, pp. 902 – 912 (1995).

[3] M. A. van den Brink, J. M. D. Stoeldraijer, and H. F. D. Linders, "Overlay and Field by Field Leveling in Wafer Steppers Using an Advanced Metrology System," SPIE Vol. 1673, pp. 330 – 344 (1992).

[4] D. S. Perloff, "A Four Point Electrical Measurement Technique for Characterizing Mask Superposition Errors on Semiconductor Wafers," IEEE Sol. St. Circ., Vol. SC-13, No. 4, pp. 436 - 444 (1978).

[5] T. Saito, S. Sakamoto, K. Okuma, H. Fukumoto, and Y. Okuda, "Mask Overlay Scaling Error Caused by Exposure Energy Using a Stepper," SPIE Vol. 1926, pp. 440 – 449 (1993).

[6] J. D. Armitage, Jr. and J. P. Kirk, "Analysis of Overlay Distortion Patterns," SPIE Vol. 921, pp. 207 – 222 (1988).

[7] T. E. Zavecz, "Lithographic Overlay Measurement Precision and Calibration and Their Effect on Pattern Registration Optimization," SPIE Vol. 1673, pp. 191 – 202 (1992).

[8] V. Nagaswami and W. Geerts, "Overlay Control in Submicron Environment," Proceedings of the KTI Microelectronics Seminar, pp. 89 – 106 (1989).

[9] M. A. van den Brink, C. G. M. de Mol, and R. A. George, "Matching Performance for Multiple Wafer Steppers Using an Advanced Metrology Procedure," SPIE Vol. 921, pp. 180 – 197 (1988).

[10] J. G. Maltabes, M. C. Hakey, and A. L. Levine, "Cost/Benefit Analysis of Mix-and-Match Lithography for Production of Half-Micron Devices," SPIE Vol. 1927, pp. 814 – 826 (1993).

[11] S. K. Jones, E. S. Capsuto, B. W. Dudley, C. R. Peters and G. C. Escher, "Wavelength Tuning for Optimization of Deep UV Excimer Laser Performance," SPIE Vol. 1674, pp. 499 – 508 (1992).

[12] M. E Preil and W. H. Arnold, "Aerial Image Formation with a KrF Excimer Laser Stepper," *Polymer Engr. and Sci.*, Vol. 32(21), pp. 1583 – 1588 (1992).

[13] R. K. Brimacombe, T. J. McKee, E. D. Mortimer, B. Norris, J. Reid, and T. A. Znotins, "Performance Characteristics of a Narrow Band Industrial Excimer Laser," SPIE Vol. 1088, 416 – 422 (1989).

[14] H. J. Levinson and R. Rice, "Overlay Tolerances for VLSI Using Wafer Steppers," SPIE Vol. 922, pp. 82 – 93 (1988).

[15] M. A. van den Brink, C. G. M. de Mol, and J. M. D. Stoeldraijer, "Matching of Multiple Wafer Steppers for 0.35 μm Lithography, Using Advanced Optimization Schemes," SPIE Vol. 1926, pp. 188 – 207 (1993).

[16] J. C. Pelligrini, "Comparisons of Six Different Intrafield Control Paradigms in an Advanced Mix-and-Match Environment," SPIE Vol. 3050, pp. 398 – 406 (1997).

[17] H. J. Levinson, M. E. Preil, and P. J. Lord, "Minimization of Total Overlay Errors on Product Wafers Using an Advanced Optimization Scheme," SPIE Vol. 3051, pp. 362 – 373 (1997).

[18] N. R. Farrar, "Effect of Off-axis Illumination on Stepper Overlay," SPIE Vol. 2439, pp. 273 – 280 (1995).

[19] T. Saito, H. Watanabe, and Y. Okuda, "Effect of Variable Sigma Aperture on Lens Distortion and Its Pattern Size Dependence," SPIE Vol. 2725, pp. 414 – 423 (1996).

[20] A. M. Davis, T. Dooly, and J. R. Johnson, "Impact of Level Specific Illumination Conditions on Overlay," Proceedings of the Olin Microlithography Seminar, pp. 1 – 16 (1997).

[21] C. S. Lee, J. S. Kim, I. B. Hur, Y. M. Ham, S. H. Choi, Y. S. Seo, and S. M. Ashkenaz, "Overlay and Lens Distortion in a Modified Illumination Stepper," SPIE Vol. 2197, pp. 2 – 8 (1994).

[22] C-M. Lim, K-S. Kwon, D. Yim, D-H. Son, H-S. Kim, and K-H. Baik, "Analysis of Nonlinear Overlay Errors by Aperture Mixing Related to Pattern Asymmetry,' SPIE Vol. 3051, pp. 106 – 115 (1997).

[23] T. A. Brunner, "Impact of Lens Aberrations on Optical Lithography," *IBM J. Res. Develop.*, Vol. 41(1/2), pp. 57 – 67 (1997).

[24] M. E Preil, T. Manchester and A. Minvielle, "Minimization of Total Overlay Errors when Matching Non-concentric Exposure Fields, SPIE Vol. 2197, pp. 753 – 769 (1994).

[25] W. W. Flack, G. E. Flores, J. C. Pellegrini, and M. Merrill, "An Optimized Registration Model for 2:1 Stepper Field Matching," SPIE Vol. 2197, pp. 733 – 752 (1994).

[26] R. Rogoff, S. S. Hong, D. Schramm, and G. Espin, "Reticle Specific Compensations to Meet Production Overlay Requirements for 64 Mb and Beyond," SPIE Vol. 2197, pp. 781 – 790 (1994).

[27] G. Rivera and P. Canestrari, "Process Induced Wafer Distortion: Measurement and Effect on Overlay in Stepper Based Advanced Lithography," SPIE Vol. 807, pp. 806 – 813 (1993).

[28] H. Izawa, K. Kakai, and M. Seki, "Fully Automatic Measuring System for Submicron Lithography," SPIE Vol. 1261, pp. 470 – 481 (1990).

[29] A. E. H. Love, *A Treatise on the Mathematical Theory of Elasticity*, 4th Edition, Dove, New York (1944).

[30] K. Chivers, "A Modified Photoresist Spin Process for a Field-by-Field Alignment System," Proceedings of the Kodak Microelectronics Seminar, pp. 44 – 51 (1984).

[31] L. M. Manske and D. B. Graves, "Origins of Asymmetry in Spin-Cast Films Over Topography," SPIE Vol. 1463, pp. 414 – 422 (1991).

[32] S. Kuniyoshi, T. Terasawa, T. Kurosaki, and T. Kimura, "Contrast Improvement of Alignment Signals from Resist Coated Patterns," *J. Vac. Sci. Technol.*, B5(2), pp. 555 – 560 (1987).

[33] G. Flores and W. W. Flack, "Photoresist Thin-Film Effects on Alignment Process Capability," SPIE Vol. 1927, pp. 367 – 380 (1993).

[34] N. Bobroff and A. Rosenbluth, "Alignment Errors from Resist Coating Topography," *J. Vac. Sci. Technol.*, B6(1), pp. 403 – 408 (1988).

[35] R. Mohondro, S. Bachman, T. Kinney, G. Meissner, and D. Peters, "High Contrast Eduction Spin Coat Process Effects on Uniformity and Overlay Registration," Proceedings of the Olin Microlithography Seminar, pp. 131 – 140 (1995).

[36] C. P. Kirk, "Theoretical Models for the Optical Alignment of Wafer Steppers," SPIE Vol. 772, pp. 134 – 141 (1987).

[37] K. Ota, N. Magome, and K. Nishi, "New Alignment Sensors for Wafer Stepper," SPIE Vol. 1463, pp. 304 – 314 (1991).

[38] D. L. Meunier, D. Humphrey, B. Peck, P. Feeney, J. Paquette, and J. Thibault, "Optimization of Metal Layer Overlay and Alignment Targets in a Chemical-Mechanical Polishing Environment," Proceedings of the Olin Microlithography Seminar, pp. 355 – 366 (1996).

[39] S-W. Hsia, G. Miyagi, and M. Brongo, "Alignment Characterization of CMP Tungsten Process," Proceedings of the Olin Microlithography Seminar, pp. 381 – 390 (1996).

[40] D. L. Meunier, B. Plambeck, P. Lord, and N. Knoll, "The Implementation of Coherence Probe Microscopy in a Process Using Chemical Mechanical Polishing," Proceedings of the OCG Microlithography Seminar, pp. 155 – 169 (1995).

[41] N. Shirishi, A. Sugaya, and D. P. Coon, "Alignment Strategies for Planarizing Technologies," SPIE Vol. 3051, pp. 836 – 845 (1997).

[42] T. Kanda, K. Mishima, E. Murakami, and H. Ina, "Alignment Sensor Corrections for Tool Induced Shift (TIS)," SPIE Vol. 3051, pp. 846 – 855 (1997).

[43] W. H. Arnold, "Overlay Simulator for Wafer Steppers," SPIE Vol. 922, pp. 94 – 105 (1988).

[44] M. H. DeGroot, *Probability and Statistics*, Addison-Wesley Publishing Company, Menlo Park, CA (1975).

[45] B. Rangarajan, M. Templeton, L. Capodieci, R. Subramanian, and A. Scranton, "Optimal Sampling Strategies for Sub-100 nm Overlay," SPIE Vol. 3332, pp.   348 – 359 (1998).

[46] G. E. Flores, W. W. Flack, S. Avlakeotes, and B. Martin, "Process Control of Stepper Overlay Using Multivariate Techniques," Proceedings of the OCG Microlithography Seminar, pp. 201 – 215 (1995).

[47] V. Barnett and T. Lewis, *Outliers in Statistical Data*, 3rd Edition, John Wiley and Sons, Chichester (1995).

CHAPTER 6
YIELD

6.1 YIELD MONITOR STRATEGY

Productivity is closely related to yield, the fraction of dies on the wafer that are functional at the end of the manufacturing process. Yield loss results from:

- Parameters that are out of specifications or improperly specified.
- Process or equipment failures.
- Chemical contamination.
- Particulate defects.

The issues of parameters such as overlay and linewidths were discussed in the prior two chapters. Process or equipment failures, which might occur, for example, when a resist pump malfunctions and wafers are not completely covered with resist, are more properly the subjects of a book on equipment maintenance, and will not be covered in this text. Chemical contamination typically results from contaminated photochemicals, and will not be considered in detail. Particulate contamination is the primary source of yield loss that lithography engineers need to address and will be the subject of this chapter.

Yield is a significant issue in lithography because of the number of masking steps required to fabricate an integrated circuit. Suppose there is a 3% yield loss at each masking step, due to random defects. While this may appear to be a small number, the total yield loss from lithography, in a 30 masking step process, is

$$\text{Yield loss} = 0.97^{30} = 0.40. \qquad (6.1)$$

A small yield loss per masking step causes an overall large yield loss, because of the large number of lithography operations involved.

Not all defects are caused by lithography:

$$\text{Yield} = LDLY \times NLLY \times PALY \times DLY \qquad (6.2)$$

where

    $LDLY$ = lithography defect limited yield,
    $NLLY$ = non-lithography limited yield,
    $PALY$ = parameter limited yield, and
    DLY = design limited yield.

All of these yield quantities are fractions between zero and one. Design limited yield occurs when parts are not designed to always yield even when all parameters are within specifications. Lithographers need to be able to measure the lithography contribution to yield, separate from other possible confounding factors.

It is also important that defects be detected before parts reach electrical test. A high level of defects at gate mask could go undetected for weeks if one relied only on electrical testing of fully fabricated devices, during which time millions of dollars of product could be lost. Electrically measurable structures,[1] such as those shown in Fig. 6.1, can be created with only a few process steps, reducing the time required for detection. For these types of monitors, an electrically conducting film deposited on an oxide-covered wafer is patterned lithographically and etched. There should be nearly infinite electrical resistance between points A and B, in the absence of defects, while the resistance between points C and D should be very low. Differences from these expected resistances indicate the existence of defects. For these types of monitors, defects may arise from films and etch, as well as lithography, and the particular process responsible for shorts or opens must still be determined after a defect is found electrically. It is also essential to have a method that can quickly assess the level of defects in particular processes, in order to see if corrective action was successful. For all of these reasons, yield engineering requires systems that are capable of inspecting wafers and identifying defects.

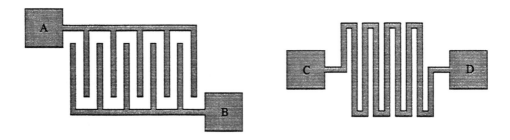

Figure 6.1. Structures for measuring defects electrically. The resistance between pads A and B should normally be very high and can provide a measurement of defects that could have been caused by particles bridging lines of resist. The resistance between pads C and D normally should be very low and can provide a measurement of defects which could have been caused by breaks in the line of resist.

Inspection systems are classified according to whether they are designed to measure patterned or unpatterned wafers. The most commonly used systems for inspecting pattern wafers are optical comparators, which compare optical signals from equivalent positions in different dies.[2] If the signals from the different positions are different, one can assume that there is a defect present at one of the positions. The optical signal is measured in a third location in order to determine which position has the defect. Holographic systems have also been used for measuring defects on patterned wafers,[3,4,5] but these systems are used less frequently than optical comparators.

Laser light scattering is used to detect defects on unpatterned wafers (Fig. 6.2). If the film shown in Fig. 6.2 were homogeneous, the light would consist only of the incident and specularly reflected beams. Particulates and bubbles result in scattered light. The physics of laser light scattering defect detection systems will be discussed in more detail in a later chapter.

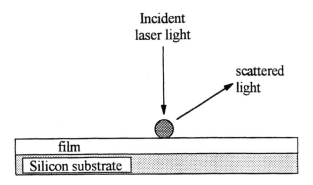

Figure 6.2. Configuration for defect detection systems using the scattering of light.

One of the biggest drawbacks of optical systems is their limited resolution. Defects as small as 100 nm can cause yield loss in 250 nm technologies, and this size of defect is falling below the detection limit of optical inspection systems. A defect inspection tool based upon a scanning electron microscope has been introduced recently, but it can only inspect for 0.1 μm defects at the rate of 0.033 cm^2/minute.[6] At this rate it is impractical to perform routine defect monitors. This is a problem that will have to be addressed in order to implement future technologies.

The following questions need to be answered when setting up systems for controlling yield:

1) Which processes will be monitored?

2) Which monitors will involve patterned wafers and which will use unpatterned wafers?

3) How many wafers should be measured?

4) Should product wafers be monitored for defects or should one use special defect monitors?

Each of these questions will be discussed in this chapter.

Lithography defects are caused almost exclusively by resist processes, which include the spin coating of photoresists, anti-reflection coatings, and developing. Particulates can be caused by mechanical systems, but these are typically very clean and cause high levels of defects only when broken. While this does happen on occasion, defects are caused far more frequently by the resist processes. Consequently, defect monitors should focus on resist coaters and developers.

Resist and anti-reflection coatings can be monitored using laser scattering systems. These have the advantage of speed, since wafers can be measured in only a few minutes, and the systems for inspecting unpatterned wafers are usually much less expensive than those that can inspect patterned wafers. If resist or anti-reflection coatings are known to have defects often, frequent monitors of these coatings can be instituted relatively easily. A defect found in early DUV processes involved a surface energy mismatch between the photoresist and the underlying anti-reflection coatings, which resulted in microbubble formation between the two films.[7,8] It was possible to detect, monitor, and control these defects using inspection systems on unpatterned wafers.

Resist patterns are usually required to identify developer defects. Typical defects associated with the develop process are residues or incomplete develop, such as missing contacts. There has recently been considerable focus on the develop process as the source of lithography defects. Systems capable of inspecting patterned wafers are usually required to identify defects which result from the develop process.

Truly random defects will appear as a Poisson process, in which the probability that a wafer will have N defects is given by

$$p(N) = \frac{d^N}{N!} e^{-d},$$
(6.3)

where d is the average number of defects per wafer. The resulting probability distribution is shown in Fig. 6.3. Control charts can be generated, but care must be taken to take possible non-normality into account. In Fig. 6.3, skewness is evident. A sufficient number of wafers must be measured to have statistical significance, and this can be a challenge when there are only a few defects per wafer.

The yield monitoring strategy will depend on process maturity. For immature processes, frequent monitors of equipment and individual processes should be used, often using unpatterned wafers.[9] This is particularly true when non-lithography processes produce high levels of defects, or when the underlying films are very grainy. Monitors that involve the use of bare silicon wafers permit smaller detection limits without overloading the system with signals from substrate defects and grains. A general recommendation is that defect monitor wafers should have 1/4 of the defects of the control limit in order to obtain good results. For example, if the control limit is 20, monitor wafers should have fewer than five defects. This is often very difficult to find on product wafers that have gone through several processes. The use of non-product wafers for defect monitoring parallels the use of non-product wafers for monitoring linewidth and overlay control that was discussed in Chapter 3.

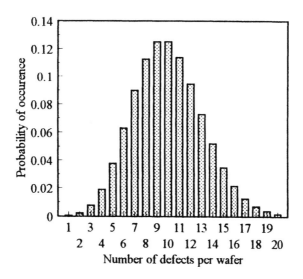

Figure 6.3. Probability of having a certain number of defects/wafer for a process where the defects are distributed randomly and the average number of defects per wafer is 10.

For mature processes, one can measure defects directly on product wafers, using automated inspection equipment. It is important to check the tools, not the lots, since the tools are the sources of defects. In general, this implies that far fewer than 100% of the lots need to be inspected. In some instances the resist processing equipment will have multiple modules per operation, in order to maintain throughput. For example, resist tracks often have two or three developer modules and two or three coater modules. In these situations the sampling plan must be designed to include wafers from every module. Scheduling measurements to properly sample tools and modules is more difficult with production wafers than with separate defect monitors. It is easier to design defect monitors which use non-product wafers. Sensitivity to very small defects is usually highest on bare silicon wafers, because of the low background of defects and substrate graininess. Where very small defects are a concern, bare silicon wafers should be used for defect monitoring. Measurement of product wafers has the advantage that additional processing time is not required, thereby maximizing photocluster availability for production, and may be used for this reason.

Defect classification provides a means for improving control of defects. The defects created by lithography processes often have signatures in their images.[10] Monitoring each defect type individually increases sensitivity for detection. For example, suppose that each defect type contributes 10 defects on average per wafer, and with a standard deviation of nine defects, wafer-to-wafer, for each of three particular defect types. Then the total defect level is 30 defects average per wafer, and the wafer-to-wafer variation is $\sigma = 16$. Suppose that the defect level for one particular type of defect increases to 15 defects average per wafer, and the standard deviation increases to 12. This is a mean shift of 0.56σ and an increase in σ of 33% for that particular defect. Overall, the mean total defect level will shift only 0.31σ, and σ overall changes by only 6%. The monitors of the individual defects are therefore more sensitive than monitors of the total defect level. There are a number of defect analysis systems that can perform defect classification automatically. Some of these are integrated into the defect detection system, while others are available on stand-alone defect review stations. Automatic defect classification not only improves sensitivity, but it also facilitates the corrective action, since the particular source of the defects is often immediately identified.

6.2 YIELD MODELS

Yield models are very important when more than one product, process or technology is involved. Yield models relate measured defect densities to product yield. It is then possible to compare yield between products to see if there is a design-related issue limiting yield on a particular product. Yield can be trended as the product mix evolves. Yield models also provide the means for extrapolating yields to new die sizes and new products yet to be produced. Most importantly, yield models can be used to facilitate improvement.[11]

Lithography defects do not always result in non-functional dies. A protrusion off a metal line will generally cause a functional failure only when the protrusion is large enough to result in bridging to another metal line. Some pieces of metal might be located in a field area where there are no relevant devices, so no damage is done. The probability k that a physical defect will result in a non-functional device is called the "kill ratio."

If we assume that the defects are random, that is, the probability of defect formation is constant in time, then the yield Y is given by

$$Y = e^{-kDA}, \qquad (6.4)$$

where D is the defect density in units of defects per unit area, and A is the die area. In addition to random defects there may be systematic yield losses. For example, products that are very sensitive to gate length may have yield loss non-randomly because of inadequate linewidth control. In the presence of systematic yield loss, the expression for yield is given by:

$$Y = Y_s e^{-kDA}, \qquad (6.5)$$

where Y_s represents the systematic yield loss.[12]

Defects often form in clusters. In this case the probability of defect formation depends on the existing level of defects. In this situation the yield follows a slightly different model:

$$Y = Y_s\left(1 + \frac{kDA}{\alpha}\right)^{-\alpha} \tag{6.6}$$

Note that Eq. 6.6 reduces to Eq. 6.5 as $\alpha \to \infty$. An example of cluster defects is shown in Fig. 6.4. These spoke defects were a consequence of a surface energy mismatch between the resist as it was being applied and an organic bottom anti-reflection coating. This mismatch resulted in the formation of microbubbles. As the wafer was spun the bubbles broke up and formed a string of smaller bubbles.

If defects are random, then yield should behave exponentially as a function of chip size. To determine the parameters of yield models with additional parameters, such as systematic yield factors or cluster parameters, data on more than one die size are needed. Often one does not have access to yield data for more than one device, so the "window method" may be used to determine whether there are significant levels of cluster defects and to calculate the cluster parameter α if the cluster model (Eq. 6.6) is appropriate.[13] In this method, yield is determined as a function of chip multiples. On the wafer layout, chips are grouped together, to form effectively larger "chips." If any real chip within the larger window has a defect, then the windowed chip is considered to have a defect.

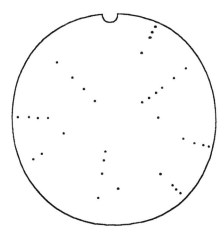

Figure 6.4. Spoke defects, generated by bubble formation at the interface between the resist and the anti-reflection coating. As the resist is spun, the bubbles break up into smaller multiple microbubbles.

The effect of clustered defects can be seen in Table 6.1. The random and cluster models do not produce significant differences in yield for small die sizes or very low defect densities. However, the random model overestimates the yield losses for larger dies.

Die size (cm²)	$D = 0.1$ defects/cm², $\alpha = 1.0$		$D = 0.3$ defects/cm², $\alpha = 1.0$	
	Random model	Cluster model	Random model	Cluster model
0.5	0.95	0.95	0.86	0.87
1	0.90	0.91	0.74	0.77
1.5	0.86	0.87	0.64	0.69
2	0.82	0.83	0.55	0.63
2.5	0.78	0.80	0.47	0.57
3	0.74	0.77	0.41	0.53
3.5	0.70	0.74	0.35	0.49
4	0.67	0.71	0.30	0.45

Table 6.1. The effect of clustered defects.

Defect distributions are frequently very non-normal. The number of defects per wafer is always non-negative, creating a problem for the lower control limit. For truly random defects, Eq. 6.3 provides the basis for determining control limits, based on a user-selected rate for false alarms. However, defects in lithography are often very non-random. Clustered defects—such as those just described—or scratches, are common, and these result in very long "tails" to the defect distributions.[14] Data are sometimes fit to lognormal or other distributions,[15,16] or transformed to a normal distribution,[17] from which control charts can be generated.

Another way to reduce sensitivity to clustered defects is to monitor the number of defective dies rather than the total number of defects, since clusters of defects are often contained within a single die. This has the additional advantage of transforming the problem to one in which conventional np control charts can be used[18] [Chapter 1, Ref. 3]. In this case p is the probability of a single die having at least one defect, and n is the number of dies measured each time.

6.3 PARAMETERS WHICH AFFECT YIELD

Lithography defects are created predominately by the resist process. Coatings of resist and anti-reflection coatings, and the develop process, have potential for causing defects. Consequently, yield improvement efforts in lithography focus primarily on the resist process.

Photoresist and developers are usually filtered at the point-of-use. As might be expected, pore size is an important factor. For state-of-the-art processing, 0.1 μm pore size is typical. The interaction of the resist and developer with the filter material is important. Air bubbles will form if the fluids do not wet the filters well. These bubbles will detach from the filter surface over time, representing a source of defects.[19] Good wetting can be accomplished by "pre-wetting" with appropriately chosen solvents or by the choice of filter materials.[20]

The thin photoresist films on the wafers are formed by spinning the wafers. Excess liquid resist is spun off the wafers, and this waste resist has the potential for hitting the walls of the coater bowl and splashing back onto the wafers. A similar problem occurs when developer is spun off wafers. Splash-back can be controlled by the design of the coater or developer bowl, choice of accelerations and spin speeds, and exhaust control. High spin speeds can easily lead to splash-back.[21] Keeping the coater bowl clean also reduces the number of defects created by splash-back. Some resist coaters have automatic bowl cleaning capabilities, while manual cleans are required periodically on systems without this feature. Daily cleans have been found to lead to measurable yield improvement.[22]

Exposed resist will dissolve in basic aqueous solutions. If the pH becomes too small the resist will not stay in solution. If water is applied to a developer puddle the pH will be reduced, and some resist may fall out of solution. Reduced defects have been found by spinning the developer off before rinsing.[23]

Steppers can be the source of yield loss, and these losses will usually appear with non-random patterns, reflecting the non-random step-and-repeat or step-and-scan process. While the defects created by resist processes are typically particulate defects, yield loss caused by steppers are usually related to parametric problems, such as loss of linewidth or overlay control. It is an extremely unusual circumstance where the Class 1 (or better) environment of the stepper is degraded to the point where there are problems with particulate contamination. Other problems with steppers may occur, and these will usually be manifested in the linewidth or overlay.

Consider, for example, the stepping pattern shown in Fig. 6.5. A magnification error would affect those dies that are farthest from the center of the exposure field. The outermost dies would have lower yield than the inner dies for an overlay-sensitive process, and this pattern would repeat from stepper exposure field to exposure field. A similar yield pattern would be evident if there was a tilt in focus across the stepper field.

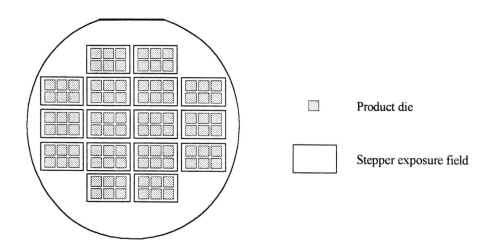

Figure 6.5. Wafer layout with multiple product dies in each stepper field.

[1] C. F. King, G. P, Gill, and M. J. Satterfield, "Electrical Defect Monitoring for Process Control," SPIE Vol. 1087, pp. 76 – 82 (1989).

[2] G. Dickerson and R. P. Wallace, "In-line Wafer Inspection using 100-Megapixel-per-Second Digital Image Processing Technology," SPIE Vol. 1464, 584 – 595 (1991).

[3] L. H. Lin, D. L. Cavan, R. B. Howe, and R. E. Graves, "A Holographic Photomask Defect Inspection System," SPIE Vol. 538, pp. 110 – 116 (1985).

[4] S. P. Billat, "Automated Defect Detection on Pattern Wafers," *Semiconductor International*, pp. 116 – 119, (May 1987).

[5] D. L. Cavan, L. H. Lin, R. B. Howe, R. E. Graves, and R. L. Fusek, "Patterned Wafer Inspection Using Laser Holography and Spatial Frequency Filtering," *J. Vac. Sci. Technol.*, B6(6), pp. 1934 – 1939 (November 1988).

[6] D. Hendricks, J. Jau, H. Dohse, A. Brodie, and D. Meisburger, "Characterization of a New Automated Electron-Beam Wafer Inspection System, SPIE Vol. 2439, pp. 174 – 183, (1995).

[7] J. Sturtevant, M. Chaara, R. Elliot, L. Hollifield, R. Soper, D. Stark, N. Thane and J. Petersen, "Antireflection Coating Process Characterization and Improvement for DUV Lithography at 0.25 μm Groundrules," SPIE Vol. 2440, pp. 582 – 593 (1995).

[8] E. H. Bokelberg and M. E. Pariseau, "Excursion Monitoring of Photolithographic Processes," Olin Microlithography Seminar, pp. 85 – 100 (1997).

[9] K. Phan, R. Chiu, S. Punjabi, and B. Singh, "Efficient and Cost Effective Photo Defect Monitoring," SPIE Vol. 3332, pp. 709 – 720 (1998).

[10] J. R. Dralla and J. C. Hoff, "Automatic Classification of Defects in Semiconductor Devices," SPIE Vol. 1261, pp. 173 – 182 (1990).

[11] D. Dance and R. Jarvis, "Using Yield Models to Accelerate Learning Curve Progress," *IEEE Trans. Semi. Manufact.*, Vol. 5(1), pp. 41 – 45 (1992).

[12] A. Y. Wong, "Statistical Micro Yield Modeling," *Semiconductor International*, pp. 239 – 148 (November, 1996).

[13] S. H. Stapper, F. M. Armstrong, and K. Saji, "Integrated Circuit Yield Statistics," Proc. IEEE, Vol. 71(4), pp. 453 – 470 (1983).

[14] D. J. Friedman and S. L. Albin, "Clustered Defects in IC Fabrication: Impact on Process Control Charts," *IEEE Trans. Semiconductor Manufacturing*, Vol. 4(1), pp. 36 – 42 (1991).

[15] D. Michelson, "A Data-Driven Method for Calculating Limits for Particle Control Charts," SPIE Vol. 2876, pp. 38 – 44 (1996).

[16] S. T. Mandraccia, G. D. Halverson, and Y-M. Chou, "Control Chart Design Strategies for Skewed Data," SPIE Vol. 2876, pp. 196 – 205 (1996).

[17] Y-M Chou, A. M. Polansky, and R. L. Mason, "Transforming Non-Normal Data to Normality in Statistical Process Control," *Journal of Quality Technology*, Vol. 30(2), pp. 133 – 141 (1998).

[18] D. Bakker and D. Icke, " Lithography Defect Characterization Using Statistical Process Control and Automated Inspection," Proceedings of the KTI Microelectronics Seminar, pp. 361 – 367 (1990).

[19] J. A. Orth, K. A. Phan, D. A. Steele, and R. Y. B. Young, "A Novel Approach for Defect Detection and Reduction Techniques for Submicron Lithography," SPIE Vol. 3050, pp. 586 – 601 (1997).

[20] M. E. Clarke and K-S. Cheng, "New Photochemical Filtration Technology for Process Improvement," Olin Microlithography Seminar poster session (1997).

[21] Y-K. Hsiao, C-H. Lee, and K-L Lu, "A Simple Method to Reduce Post Develop Residue," Olin Microlithography Seminar poster session (1997).

[22] Y-T. Fan, H-P. Lin, Y-C. Lo, C-H. Lee, and K-L. Lu, "A Study on Methods to Reduce Metal Defects Caused by a Coating Process," Olin Microlithography Seminar poster session (1997).

[23] E. H. Bokelberg, J. L. Goetz, and M. E. Pariseau, "Photocluster Defect Learning and Develop Process Optimization," Olin Microlithography Seminar, pp. 127 – 139 (1996).

CHAPTER 7
PROCESS DRIFT AND AUTOMATIC PROCESS CONTROL

7.1 ADJUSTING FOR PROCESS DRIFT

There are times when processes drift, and the best course of action is to make an adjustment. This situation occurs when overlay correctable parameters or focus offsets are changed. Every lithography operation should have controls to detect process drift. As discussed in Chapter 1, the ±3σ rule is not particularly sensitive to shifts in the process mean, compared to other Western Electric Rules. Consequently, it is advisable to include at least one additional Western Electric Rule to supplement the ±3σ rule to increase sensitivity to process drift. An alternative to the Shewhart control chart, with greater sensitivity to shifts in the process mean, is the exponentially-weighted moving average (EWMA) chart, discussed in the next section.

Suppose that a process has been under control until a Western Electric Rule is violated. If, after investigation, no assignable cause can be identified, it may be decided that the process should be adjusted. Determining the amount of adjustment is the subject of this section. Each of the Western Electric Rules involves a certain amount of data. The ±3σ rule is violated when a single point exceeds control limits, while the eight-in-a-row rule involves 8× more pieces of data. As discussed earlier, the process mean μ is known to within a "±3σ" level of confidence (99.7%) following n measurements:

$$\mu = \bar{x} \pm \frac{3\sigma}{\sqrt{n}} \qquad (7.1)$$

With more measurements there will be a smaller uncertainty in the actual process mean (Fig. 7.1). For this reason, it is useful to repeat measurements when Western Electric Rules have been violated. This is particularly true for the ±3σ rule, for which there may be only a single measurement indicating a loss of process control. As seen in Fig. 7.1, the violation of the ±3σ rule could have been caused by only a very small shift in the process mean. Changing the operating point to move the single violation to the process mean would most likely represent over-compensation. The exponentially weighted moving average, discussed in the next section, provides a means of both detecting and estimating the process mean, particularly in situations involving gradual drift.

Suppose only the ±3σ rule has been violated, and it is proposed to adjust the process assuming that the single measurement represents a new shifted process. The process is then adjusted on the basis of this assumption to some percentage of the difference between the single value and the process target.

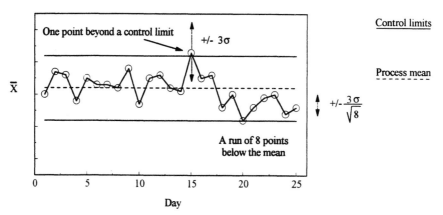

Figure 7.1. **X-bar chart, where the uncertainty in the process mean is greater when based upon a single value, compared to eight values.**

The operation is performed again and new measurements are taken, and the process is again adjusted on the basis of a single determination of the state of the process. Ultimately the mean of the process will converge to the desired target. It has been shown that the most efficient way to conduct this procedure is to correct for 100% of the difference between the measured value and the target on the first iteration, 50% on the second, and $100 \times \frac{1}{n}\%$ on the nth iteration.[1] The final result is identical to that obtained by performing the operation n times, taking the measurements, and making a single adjustment, but has the advantage that an adequate process may be reached more quickly. The primary difficulty with any attempt to determine the true state of the process from multiple tests and measurements is the need to account for fluctuations which occur over time scales greater than a few hours. This can be accomplished only by monitoring the process over the appropriate time scale.

Example: Consider a process where the linewidth target is 250 nm. Suppose that, although the correct exposure dose is unknown, the exposure latitude is known to give a 25 nm shift in linewidth for a 10% change in exposure dose. Using the algorithm where the correction on the nth iteration is $100 \times \frac{1}{n}\%$, one obtains the following.

Iteration	Exposure dose	Measured mean linewidth	100% correction	$100 \times \frac{1}{n}\%$ correction
0	20.0 mj/cm^2	280 nm	2.4 mj/cm^2	2.4 mj/cm^2
1	22.4 mj/cm^2	260 nm	0.8 mj/cm^2	0.4 mj/cm^2
2	22.8 mj/cm^2	242 nm	- 0.6 mj/cm^2	- 0.2 mj/cm^2
3	22.6 mj/cm^2	248 nm		

7.2. THE EXPONENTIALLY-WEIGHTED MOVING AVERAGE

The exponentially-weighted moving average (*EWMA*) for measurement t is defined as[2]:

$$EWMA_t = \lambda y_t + (1-\lambda)EWMA_{t-1} \tag{7.2}$$

$$= \lambda y_t + \lambda(1-\lambda)y_{t-1} + \lambda(1-\lambda)^2 y_{t-2} + \cdots \tag{7.3}$$

where $0 < \lambda \leq 1$ and $t = 1,2,...$, and y_t is the parameter of interest. To calculate $EWMA_1$ the process target or some prior estimate of the process mean for "$EWMA_0$" can be used. The exponentially-weighted moving average provides a measure of the process mean with more recent data being weighted more heavily than older measurements. Smaller values of λ weight older data more heavily than larger values. As $\lambda \rightarrow 1$, the EWMA weighs the most recent data more heavily, and control charts of the exponentially-weighted moving average begin to resemble a standard Shewhart chart, such as shown in Fig. 1.5. The value of λ is something that must be determined by the person setting up the control chart. Considerations needed to set this value will be discussed shortly.

If the y_t are independent normally distributed random variables with standard deviation σ, then the standard deviation of the exponentially-weighted moving average is known in closed form[3]:

$$\sigma_{EWMA_t} = \sigma\sqrt{\frac{\lambda}{2-\lambda}\left[1-(1-\lambda)^{2t}\right]} \tag{7.4}$$

$$\rightarrow \sigma\sqrt{\frac{\lambda}{2-\lambda}} \quad \text{for large values of } t. \tag{7.5}$$

From this, control limits for an exponentially-weighted moving average control chart can be generated:

$$\text{Upper control limit} = \mu + L\sigma\sqrt{\frac{\lambda}{2-\lambda}\left[1-(1-\lambda)^{2t}\right]}, \tag{7.6}$$

$$\text{Lower control limit} = \mu - L\sigma\sqrt{\frac{\lambda}{2-\lambda}\left[1-(1-\lambda)^{2t}\right]}, \tag{7.7}$$

where μ is the process center line and L is another parameter to be determined by the person setting up the control chart. These control limits approach constant values as t gets large.

The values for λ and L can be determined once two questions have been answered:

1) What is the desired average run length (ARL) for the process when it is in control?

2) What is the minimum shift at which detection is desired?

Values which lead to optimized EWMA control charts have been calculated for various values of the in-control ARL and process shifts, for normal and exponentially distributed processes. Examples of values for λ and L are given in Table 7.1 for a normally distributed process.[4]

Minimum shift (in units of σ)		In-control average run length		
		100	300	500
0.5	λ	0.065	0.055	0.050
	L	1.98	2.43	2.62
	ARL at shift	17.3	24.9	28.7
1.0	λ	0.175	0.145	0.135
	L	2.32	2.71	2.71
	ARL at shift	7.0	9.1	10.2

Table 7.1. Values for λ and L for establishing EWMA control charts for minimum detectable shifts of 0.5 and 1.0 σ, for a various average run lengths for the in-control process.

EWMA control charts are sensitive to small shifts in the process mean. As can be seen from the data in Table 7.1 and Fig. 1.3, this sensitivity is comparable to conventional Shewhart charts used in conjunction with the Western Electric Rules, but with fewer false alarms for processes in control.[5] The EWMA charts also have the advantage of providing the user with an immediate estimate of the process mean. As seen in the preceding section, this is not an immediate result of the Shewhart control charts. This is illustrated by an example. Consider the linewidth data shown in Fig. 7.2, for a process with an in-control mean of 250 nm and control limits of ± 25 nm (σ = 8.33 nm) about the mean. After 50 measurements, drift was introduced into the process, and the linewidths grew larger, on average, over time. After 104 data points the "eight-in-a-row" Western Electric Rule was violated, and the "± 3σ" rule was violated at the 115th point. The corresponding EWMA average control chart for this process is shown in Fig. 7.3. Control limits were calculated to provide sensitivity to small shifts and were assumed to be based on a large number of prior data (t large). An in-control average run length of 300 and a minimum detected shift of 0.5σ were chosen, giving a value of 0.055 for λ (Table 7.1). From Eqs. 7.6 and 7.7, the control limits for the EWMA control chart are 250 ± 3.41 nm. The drift in the process was detected at point 77 with the EWMA control chart. A trend upwards in the mean of the process can be seen in both charts, but it is more evident in the EWMA control chart. The EWMA is particularly useful for calculating process parameters in the feedback systems discussed in Chapter 3.

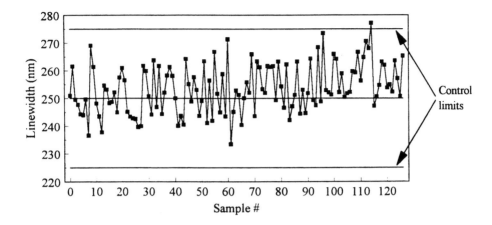

Figure 7.2. A Shewhart variables control chart for a process that experienced a drift in the process after sample #50.

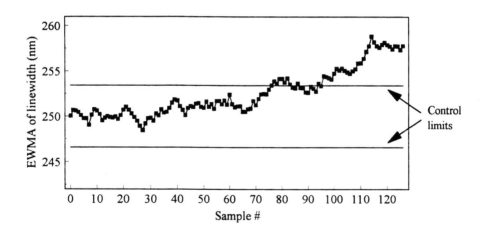

Figure 7.3 The EWMA control chart for the same process illustrated in Figure 7.2.

7.3. AUTOMATIC PROCESS CONTROL

With an estimate of the process mean, a target value, and a rule for making adjustments, one can consider automatic process control. The equipment used in lithography utilizes a considerable amount of automatic process control. Consider, for example, a hotplate used for resist processing. The temperature of the hotplate is measured with an appropriate temperature probe, and additional power is provided to the heating elements if the temperature is too low, or the power is reduced if the temperature is too hot. There are many other process factors which are controlled automatically by the equipment, including exposure dose, focus, overlay, coater spin speeds, and exhausts.

It is common practice in some industries to adjust the process based on critical product characteristics. For example, if one is making a batch of resist developer, more water would be added if the normality were too high, or more base might be added if the normality were too low. This adjustment of developer normality is possible because the critical parameter — normality — can be measured and changed during processing. On the other hand, one cannot directly measure parameters that are critical to lithography, such as linewidth and overlay, and make corrections during processing. The necessary measurements require highly specialized equipment. One cannot perform a scanning electron microscope (SEM) measurement of linewidths while wafers are being exposed or developed.

It has been proven that it is possible to monitor the lithography process indirectly, and thereby improve process control. The most notable technique for doing this monitoring is develop end-point detection. In develop end-point detection the thickness of the resist is monitored during develop. This can be accomplished using optical thin film interference. This technique was developed originally for characterizing resist development in a laboratory,[6] and was later extended to real-time application.[7, 8] In this method, light is incident on the resist coated substrate, and the intensity of reflected light is detected. During development of the resist the optical signal will vary because of thin film interference effects (Fig. 7.4). Eventually the resist is developed completely to the substrate in large open areas (point C). This signals the transition of one part of the development process to another. If the develop time between point C and the desired end of the process is fairly constant, then most of the variations in the process can be captured in the variations in the amount of time for the develop cycle to reach point C. A system that can determine the time when the resist has just been developed in the large open areas (point C) can be used to control the process automatically.

In some ways this approach is very attractive. Process control can be accomplished by equipment in real time. Numerous control charts do not need to be maintained, and people are not needed to review the control charts and adjust processes. However, there is no such thing as a free lunch. Develop end-point controllers require some teaching, as every process behaves somewhat different from the others. Some layers, such as contacts, often provide interference signals that are too weak for develop endpoint detection to be useable. Moreover, develop endpoint controllers are not available from the same equipment suppliers who make the rest of the resist processing equipment, and integration is often problematic. This approach is also difficult to apply to multiple-puddle processes, which are often implemented because of overriding concerns about yield. Finally, many resists for DUV lithography develop very quickly, taking only a few seconds. This is too fast for effective develop end-point control. For these reasons, develop endpoint detection has never been used extensively, although it has been demonstrated to improve linewidth control.[9] Develop monitors have not found widespread acceptance for in-line use, but they have been essential tools for determining parameters for image modeling[10] and in-line[11] and off-line[12] process monitoring.

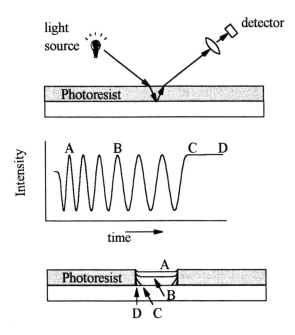

Figure 7.4. The develop end-point detection system.

Recently, the use of latent images for process monitoring has been explored.[13] A latent image is the pattern in the resist film, prior to development, that results from the difference in optical constants or film thickness between exposed and unexposed resist. For example, the latent image can be monitored during post-exposure bake of chemically-amplified resists. Feedback could be applied to the post-exposure bake time.[14, 15, 16] With i-line resists the information gathered while measuring the latent image can be fed forward to the develop step. So far, these techniques have remained primarily in the laboratory,[17, 18, 19, 20, 21, 22] for a number of reasons. A significant complication of these techniques is the non-equivalence between resist development or post-exposure bake and other factors which affect linewidth, when many different feature types are considered. For example, if there is defocus, the focus should be corrected, rather than having the develop or bake time varied.

Among those who study process control, the nature of automatic process control and its relationship to statistical process control have become subjects of considerable interest.[23, 24, 25] Automatic process control can be applied when the state of a process can be measured, and the process can be regulated through a control algorithm on the basis of these measurements. The other extreme situation is exemplified by Deming's funnel experiment, in which the process is completely random and not amenable to any control. Most processes involve some potential for regulation as well as a significant level of statistical noise that must be addressed.

Ultimately, the problem for applying automatic process control to lithography remains one of metrology. Obtaining good data on which to make decisions is critical. As will be seen in the next chapter, when linewidths and overlay are measured in nanometers, this is difficult even under conditions optimized for the purpose of making measurements.

---

[1] F. E. Grubbs, "An Optimum Procedure for Setting Machines or Adjusting Processes," *Journal of Quality Technology*, Vol. 15(4), pp. 186 – 189 (1983).

[2] S. V. Crowder, "Design of Exponentially Weighted Moving Average Schemes," *Journal of Quality Technology*, Vol. 21, pp. 155 – 162 (1989).

[3] J. S. Hunter, "The Exponentially Weighted Moving Average," *Journal of Quality Technology*, Vol. 18, pp. 203 – 210 (1986).

[4] J. M. Lucas and M. S. Saccucci, "Exponentially Weighted Moving Average Control Schemes: Properties and Enhancements," *Technometrics*, Vol. 32, pp. 1 – 12 (1990).

[5] G. E. Flores, W. W. Flack, S. Avlakeotes, and M. Merrill, "Monitoring and Diagnostic Techniques for Control of Overlay in Steppers," SPIE Vol. 2439, pp. 40 – 60 (1995).

[6] K. L. Konnerth and F. H. Dill, "In-situ Measurement of Dielectric Thickness During Etching or Developing Processes," *IEEE Trans. Electron Devices*, Vol. ED-22, pp. 452 – 456 (1975).

[7] L. J. Lauchlan, K. M. Sautter, and T. Batchelder, "Automatic Process Control for VLSI Linewidth," *Solid State Technology*, pp. 333 – 337 (April, 1985).

[8] L. J. Lauchlan, K. Sautter, T. Batchelder, and J. Irwin, "In-Line Automatic Photoresist Process Control," SPIE Vol. 539, pp. 227 – 233 (1985).

[9] C. Nygren, J. Daggett, and J. Grambow, "The Use of Develop End Point Detection to Eliminate Photolithography Process Variation," Proceedings of the KTI Microlithography Seminar, pp. 315 – 348 (1990).

[10] P. D. Flanner III, "Improved Methods of Photoresist Development Characterization," Proceedings of the KTI Microelectronics Seminar, pp. 231 – 238 (1987).

[11] K. M. Sautter, M. Ha, and T. Batchelder, "Development Process Control and Optimization Utilizing an End Point Monitor," Proceedings of the KTI Microelectronics Seminar, pp. 99 – 112 (1988).

[12] J. A. Bruce and B. J. Lin, "Determination of Exposure Dose by Photoresist Development Rate," Proceedings of the KTI Microelectronics Seminar, pp. 1 – 11 (1987).

[13] M. L. Miller and D. A. Mellichamp, "Development of an End-Point Detection Procedure for the Post-Expose Bake Process," SPIE Vol. 2439, pp. 78 – 88 (1995).

[14] S. Zaidi, S. L. Prins, J. R. McNeil, and S. S. H. Naqvi, "Metrology Sensors for Advanced Resists," SPIE Vol. 2196, pp. 341 – 351 (1994).

[15] J. L. Sturtevant, S. J. Holmes, T. G. Van Kessel, P. D. Hobbs, J. C. Shaw, and R. R. Jackson, "Postexposure Bake as a Process-Control Parameter for Chemically Amplified Photoresist," SPIE Vol. 1926, pp 106 – 114 (1993).

[16] J. L. Sturtevant, S. Holmes, T. G. Van Kessel, M. Miller, and D. Mellichamp, "Use of Scatterometric Latent Image Detector in Closed Loop Feedback Control of Linewidth," SPIE Vol. 2196, pp. 352 – 359 (1994).

[17] K. C. Hickman, S. M. Gaspar, K. P. Bishop, S. S. H. Haqvi, J. R. McNeil, G. D. Tipton, B. R. Stallard, and B. L. Draper, "Use of Diffracted Light from Latent Images to Improve Lithography Control," SPIE Vol. 1464, pp. 245 – 251 (1991).

[18] T. E. Adams, "Applications of Latent Image Metrology in Microlithography," SPIE Vol. 1464, pp. 294 – 312 (1991).

[19] S. L. Prins and J. R. McNeil, "Scatterometric Sensor for PEB Process Control," SPIE Vol. 2725, pp. 710 – 719 (1996).

[20] L. M. Milner, K. C. Hickman, S. M. Gaspar, K. P. Bishop, S. S. H. Haqvi, J. R. McNeil, M. Blain, and B. L. Draper, "Latent Image Exposure Monitor Using Scatterometry," SPIE Vol. 1673, pp. 274 – 283 (1992).

[21] M. L. Miller and D. A. Mellichamp, "Development of an End-Point Detection Procedure for the Post-Exposure Bake Process," SPIE Vol. 2439, pp. 78 – 88 (1995).

[22] T. Koizumi, T. Matsuo, M. Endo and M. Sasago, "CD Control using Latent Image for Lithography," SPIE Vol. 2439, pp. 418 – 426 (1995).

[23] D. C. Montgomery, J. B. Keats, G. C. Runger, and W. S. Messina, "Integrating Statistical Process Control and Engineering Process Control," *Journal of Quality Technology*, Vol. 26(2) pp. 79 – 87 (1994).

[24] The April, 1997, Vol. 29(2) of the *Journal of Quality Technology* has a discussion on statistically-based process monitoring and control, with contributions from several authors and extensive references.

[25] G. Box and A. Luceño, *Statistical Control by Monitoring and Feedback Adjustment*, John Wiley & Sons, New York (1997).

CHAPTER 8
METROLOGY

. . . .

8.1 THE NEED FOR UNDERSTANDING THE MEASUREMENT PROCESS: DEFECT DETECTION

The information upon which operators, technicians and engineers make decisions consists of *measured* data. Bad data usually lead to bad decisions. For this reason quality control professionals have always focused on metrology. The problem is particularly significant for lithographers, where the extremely small features of state-of-the art processes have pushed measurement capabilities to their limits, and, some might say, beyond. Some of the metrology issues and challenges faced by lithographers will be discussed in this chapter.

Defect detection will be used as an example to illustrate the importance of understanding the measurement process in order to properly use measurement tools. Consider a laser light scattering system used to detect defects (Fig. 6.2). A particle is detected because it scatters light. However, small particles scatter light primarily in the forward direction (Fig. 8.1). When defects are sitting on, or embedded in, thin films, optical interference will affect the strength of the signal produced by the scattered light, because the forward scattered light is detected only after it has reflected from the substrate (Fig. 8.2). Since particle size is determined by the strength of the scattered light signal in these systems, the "measured" particle size will depend upon the films on the substrate, which can modulate the reflectance (Fig. 8.3). A particle will scatter a certain amount of light, but the films on the substrate will affect the amount of light that actually reaches the detector. Consider, for example, the "measured" particle size in a controlled experiment. Polystyrene latex spheres were used to calibrate a system which used a HeNe laser (wavelength = 632.8 nm) on bare silicon. The resulting measured particle sizes for spheres on bare silicon and substrates with different thickness of silicon nitride are shown in Table 8.1.[1] Because of the thin film interference effect, the apparent particle size is generally smaller for the nitride films, since a reduced signal can result from either small particles or a low-reflectance substrate. In some instances, the signal was too small to provide a measurement. The opposite effect occurs on an aluminum substrate, where the highly reflective surface appears to enhance the particle size (Table 8.2). Interpretation of defects that consist of embedded particles or voids must be made with caution, as well.

It has just been shown that considerable misinterpretation of the data would occur if one did not take thin film optical interference effects into account. This is of particular importance to lithographers, who are usually concerned about defects in photoresist films or anti-reflection coatings. Correct decisions require an understanding of the measurement process. This has been illustrated here by examining defect detection, but it is generally good engineering practice to understand all metrology processes in order to be able to properly interpret acquired data. From this perspective, the measurement of linewidths by scanning electron microscopes is discussed in the next section.

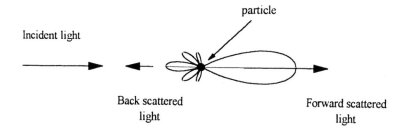

Figure 8.1. The angular distribution of light scattered from a spherical particle.

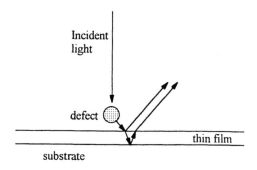

Figure 8.2. Geometry for detection of defects on thin films. Light is scattered from the defect and then reflected from the substrate. It is the reflected light which is detected. Because of interference effects the reflected signal will vary with film thickness.

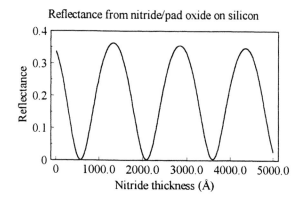

Figure 8.3. Because of interference effects between incident and reflected light, signal from particles on thin films will vary with film thickness. The calculated reflectance is for normally incident light, also collected near-normal, with a wavelength of 632.8 nm (HeNe laser).

Latex sphere diameter (μm)	Nitride thickness		
	0 Å	825 Å	1400 Å
0.36	0.26		0.19
0.50	0.53		0.55
0.62	0.59		0.61
0.76	0.71		0.73
0.90	0.74	0.084	0.83
1.09	0.94	0.13	0.96
2.02	2.1	0.21	1.50
5.00	6.7	0.75	3.23

Table 8.1. Measured particle size for polystyrene latex spheres on nitride films of different thicknesses.

Latex sphere diameter (μm)	Scattering cross section on silicon	Scattering cross section on aluminum
0.36	0.26	0.78
0.62	0.59	1.54
1.09	0.94	2.66
2.02	2.1	6.33
5.00	6.7	20.46

Table 8.2. Measured particle size for polystyrene latex spheres on silicon and aluminum surfaces.[1]

8.2 LINEWIDTH MEASUREMENT USING SCANNING ELECTRON MICROSCOPES

The most common tool for measuring linewidths is the scanning electron microscope (SEM). Low voltage SEMs are capable of measuring resist features in-line. As shown in the prior section, understanding the measurement process is essential for applying a measurement technique correctly, and this is particularly true of linewidth measurements. In the scanning electron microscope, electron beams are scanned across patterns on wafers. The voltage of the electron beams ranges between a few hundred volts to tens of thousands of volts. For measuring resist features, a typical range is 400 – 1000 volts. The incident beam is scattered, both elastically and inelastically, producing secondary and back-scattered electrons.[2,3] By commonly accepted definition, secondary electrons are those with energy less than 50 eV, while the back-scattered electrons are those with energies closer to the energy of the incident beam. The secondary electrons created by inelastic scattering or the elastically back-scattered electrons are detected synchronously with the scan of the incident beam (Fig. 8.4). Because the number and direction of the scattered electrons depends upon the material composition and topography of the features over which the incident beam is scanned, the intensity of the detected signal varies so that an image can be created.

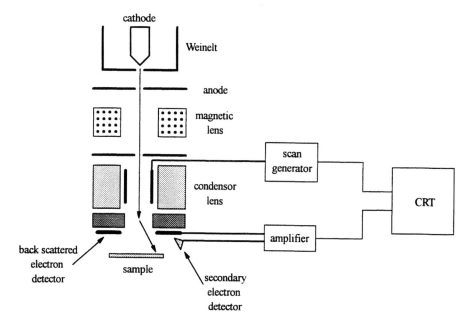

Figure 8.4. Cross-sectional view of a scanning electron microscope.

As the incident beam is scanned across a line, the detected signal varies. Algorithms are required to relate the edges of the line to points on the intensity versus position curve, and these algorithms do not always provide complete accuracy. The signal at the edge of a feature is shown in Fig. 8.5. A mathematical algorithm is required to identify a point on the signal profile to represent the edge of the line. A number of edge detection methods are used.[4] The simplest algorithm uses a threshold value for identifying this point:

$$\text{Threshold level} = (1 - P) \times \text{minimum} + (P) \times \text{maximum}, \qquad (8.1)$$

where P is a value between 0 and 1, typically in the neighborhood of 0.5. Other methods include maximum slopes, inflection points, linear approximations, minima or maxima.

Secondary and back-scattered electron emissions are proportional to slopes of features,[5] typically varying as $1/\cos(\theta)$ (Fig. 8.6). Steep slopes appear "bright" in scanning electron microscopes. Consequently, the signal depends on the slope of the feature, something which can change easily with stepper defocus and other process variations for resist features, and variations in etch processes can change the slopes of etched features. By measuring linewidths and slopes from cross sections, and comparing these results to in-line SEM measurements, a careful study found that measurements of 180 – 250 nm lines could vary by as much as 10 nm per degree change in line-edge slope[6] (Fig. 8.7).

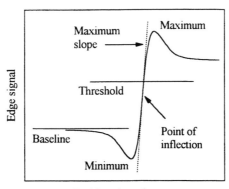

Figure 8.5. The electron signal corresponding to one edge of a feature being measured by a scanning electron microscope. Several attributes of the feature can potentially be used to determine the location of the edge of the feature.

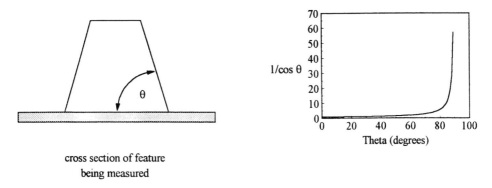

Figure 8.6. Slope of a feature being measured, and the secondary electron production as a function of this slope.

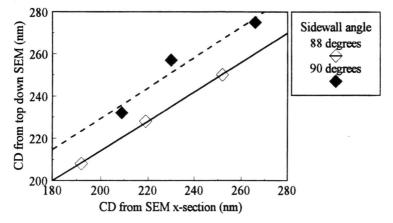

Figure 8.7. Top-down SEM measurements versus measurements from cross sections of the same etched polysilicon features (Ref. 6). Measurement error bars are not shown.

Another problem associated with SEM linewidth measurements—sample charging—is a particularly significant problem for measuring features comprised of photoresist. Consider an electron beam focused onto a homogeneous substrate. At low incident beam voltages more electrons are introduced into the sample than are emitted in the form of secondary and back-scattered electrons. As the beam voltage is increased, this balance changes, and samples acquire a net positive charge during imaging[7] (Fig. 8.8). At yet higher voltages the net charge in the sample reverses yet again. Imaging and measurement is usually most stable at the point where the samples remain electrically neutral. Point E_2 is the more stable of the two, and scanning electron microsopes are typically operated close to this voltage. For conductive materials this voltage ranges from $2 - 4$ kV, while it is only $0.5 - 1.0$ kV for non-conducting materials such as photoresist.[203] Since imaging is typically superior at higher voltages, SEM metrology for lithography is at a distinct disadvantage.

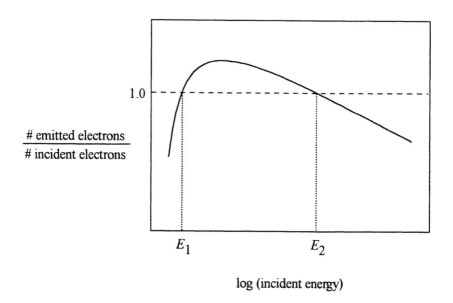

Figure 8.8. At incident energies E_1 and E_2 the sample will be electrically neutral while it will charge for other energies. The values of E_1 and E_2 are material dependent.

The charging of features measured in lithography is a complex phenomenon, and the idea of a single voltage at which there is charge neutrality is inadequate. The incident electrons and emitted secondary and back-scattered electrons are not the only sources of electrical charge that need to be considered in order to understand sample charging. Samples in scanning electron microscopes are typically grounded, and there will be a current between the surface of the sample and ground. This current will be proportional to the voltage difference between the sample surface and ground, and it will depend upon the resistivity of the materials that comprise the sample. Highly conductive materials will not charge, because there will be a steady current to ground. This charge balance can be written as:

$$I_B = (\delta + \eta)I_B + \frac{E_{surface}}{R} \tag{8.2}$$

where I_B is the incident beam current, δ is the secondary electron yield, η is the back-scattered electron yield, $E_{surface}$ is the surface potential, and R is the effective resistance to ground. Consequently, the surface potential is given by[8]:

$$E_{surface} = I_B R(1 - \delta - \eta) \tag{8.3}$$

The surface potential is seen to be proportional to beam current, once the current to ground is taken into consideration. Conductivity to ground can be complex in insulating materials, where the excitation of electrons to the conduction band will be a function of beam energy and current. Complicating the situation even more is the ability of insulators to trap charge. The time it takes to discharge the trapped charges as a consequence of thermal relaxation is on the order of minutes.

Unfortunately, this picture is still incomplete for the types of measurements that occur in lithography applications and typically involve features patterned into photoresist. As noted previously, secondary electron emission depends on the angle between the incident beam and the plane of the surface being probed. Consequently, the electron beam energy at which there is charge neutrality will vary between the top and sides of particular features, because the rate of emission of secondary and back-scattered electrons from the tops of features will differ from the rate from the sides. The photoresist is nearly always developed to the substrate in exposed areas, and the substrate will have a different value for E_2 than the photoresist.

Moreover, the surface potential may be more important than overall charge neutrality. It is the electrical potential at the surface of the features, relative to the potential at the detector, that determines the flow of secondary electrons from the object being measured to the detector. Overall charge neutrality is important for electrons far from the object, but the distribution of electrons within and near the object will determine the electrical potential at the object's surface. The issue of sample charging remains a topic of current research.

Because photoresist is not electrically conducting it is difficult to dissipate charge once accumulated. Charging also affects the linewidth measurement. A negatively charged line, which will occur for higher voltages, will deflect the electron beam and result in a narrower measurement[9] (Fig. 8.9).

Scanning beam electrons

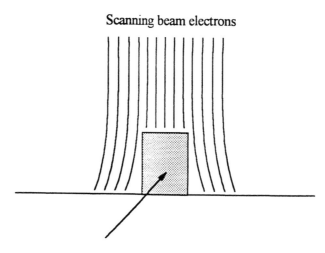

Electrically charged photoresist

Figure 8.9. Scanning beam electrons are deflected away from the negatively charged photoresist. As a consequence, the line measures narrower than it actually is.

Once secondary electrons are generated they need to escape from the sample and hit a detector in order to create a signal. SEM measurements of contact holes have been difficult for this reason, since it is difficult for low-energy secondary electrons to emerge from positively charged holes.[10] For this reason, detection of back-scattered electrons is sometimes used for metrology with lithography applications. Because back-scattered electron emission peaks at or near the direction of the incident beam, detectors are sometimes placed within the lens.[11]

The conversion of a secondary or back-scattered electron intensity into a linewidth number is not automatic. It may be argued that measurement precision is the critical requirement, and that absolute accuracy is less important. This is true to a certain extent, but facilities with more than one SEM, or those that need to transfer processes to or from others, will require standards to match SEMs. Consequently, some level of calibration is usually required, and this gauge-setting is generally material-dependent. Algorithms for calibrations on resist features will differ from those for polysilicon, and will depend on the thickness and slopes of the resist features. Consequently, there are very few standards available from organizations such as National Institute of Standards & Technology (NIST) that can be used directly in lithography applications. Current NIST linewidth standards, SRM 473, 475, and 476, are chromium-on-quartz, intended primarily for measuring features on photomasks.[12] SRM 473 has linewidths down to 0.5 μm. There have been efforts to develop standards designed specifically for use in SEMs.[13,14] An interim standard, SRM 8090, is available from NIST, consisting of palladium/titanium lines on silicon, and it has linewidths down to 0.2 μm. It is intended that this interim standard will be superseded eventually by SRM 2090. Since SEMs will "measure" different linewidths for features made from different materials or thickness, these standards are useful only for ensuring measurements on photomasks or controlling such parameters as the magnification of the SEM. However, the magnification can be verified by measuring the pitch of grating patterns,[15, 16] which will give measurements independent of materials and measurement algorithms. Although commercially available SEM-based linewidth measurement systems may not produce measurements that are accurate (i.e., give the "real" linewidth), they may be precise and repeatable. The difference between accuracy and precision is illustrated in Fig. 8.10.

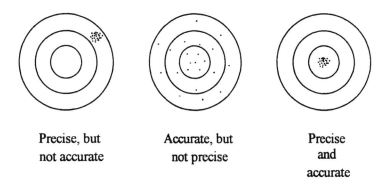

Precise, but   Accurate, but   Precise
not accurate   not precise   and
                                             accurate

Figure 8.10. Shots at a target, illustrating precision and accuracy.

One problem preventing the creation of good linewidth standards for lithography is carbon deposition during SEM measurement.[17] While the electron beam is scanning the sample, carbon is being deposited. As one might expect, this is a particularly significant problem for photoresist samples. Repeated measurements lead to changes in the measured linewidth, even though the pitch remains unchanged.[18] One approach to dealing with the change in measured linewidth due to carbon deposition on the standard is to predetermine the magnitude of deposition-induced linewidth change per measurement, and subtract that from the monitor values.[19] SPC can then be applied to the residuals.

While an SEM can reproduce the linewidth on a particular standard artifact, there may be some concern that linewidth measurements of other materials, such as photoresist, may not be properly calibrated. To this end, atomic force microscopes can be used to measure photoresist linewidths for the purpose of calibration.[20] Atomic force microscopes are purely mechanical and will not have the type of material dependence that scanning electron microscopes have.

As with any measurement system, the proper interpretation of the numbers produced by atomic force microscopes requires an understanding of how they measure. For example, because of probe tip size, atomic force microscopes will not provide reliable measurements of structures with very tight pitches. Nevertheless, the atomic force microscope is another useful metrology tool.

Calibration usually requires measurements of multiple linewidths. If r is the "real" linewidth, then the measurement linewidth y is given by:

$$y = ar + b \tag{8.4}$$

to first order, where a and b are constants. Measurement of more than one linewidth is required to determine the two calibration parameters, as well as reduce errors from measurement noise.

The SEM used for metrology must be properly focused and stigmated in order to obtain good measurements. In the past this was accomplished by using a sample of very small grains, such as sputtered gold or grainy silicon, and adjustments were made until the image appeared best to the operator. Recent work has shown that the subjective element associated with this approach can be overcome by taking a two-dimensional Fourier transform of the image.[21] Defocus will involve a decrease in the magnitudes of the higher frequency components, and astigmatism will show asymmetries between axes.

8.3 ELECTRICAL LINEWIDTH MEASUREMENT

Linewidths can be measured electrically as a supplement to SEM measurements. This method has the advantage of speed; considerable data can be collected in a small amount of time, once wafers have been patterned and etched. Off-line stepper lens setup and characterization often requires that large numbers of linewidths be measured, and the delays incurred because wafers must be etched is tolerable. For example, the determination of best focus and field tilt requires that linewidths be measured at many points in the field, with features of at least two orientations, and at several focus settings. It is useful to measure linewidths in multiple fields in order to average out noise from factors not related to the lens and optical column tilt. It is not practical to measure this many linewidths with a scanning electron microscope, even an automated one.

Electrical linewidth metrology is a practical method for this type of setup.[22] Linewidths of electrically conducting material, such as polysilicon, silicide, or metal, can be measured using structures such as those shown in Fig. 8.11.[23] The electrical linewidth W is related to two measured resistances:

$$W = \frac{R_s}{R_b} L. \tag{8.5}$$

R_s is the sheet resistance and it is determined as follows. A current I is forced between pads 3 and 4, and a voltage V_{25}^+ is measured between pads 2 and 5. The current is reversed, and a voltage V_{25}^- is measured. The current I is then forced alternately between pads 2 and 3, and voltages V_{45}^+ and V_{45}^- are measured. The sheet resistance is given by:

$$R_s = \frac{\pi}{2} \frac{\left|V_{25}^+\right| + \left|V_{25}^-\right| + \left|V_{45}^+\right| + \left|V_{45}^-\right|}{4I}. \qquad (8.6)$$

R_b is determined by forcing current I_b alternately between pads 1 and 3 and measuring the voltages between pads 5 and 6.

$$R_b = \frac{\left|V_{56}^+\right| + \left|V_{56}^-\right|}{2I_b} \qquad (8.7)$$

To avoid end effects, typically $W \ll L$.

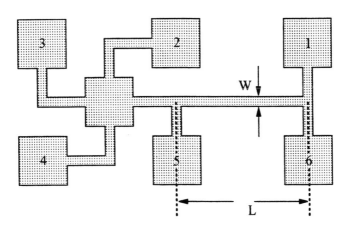

Figure 8.11. The resistor test structure for measuring linewidths electrically.

Electrically measured linewidths, as determined from Eqs. 8.5 – 8.7, tend to disagree in absolute value with measurements from other metrology methods, such as SEMs and atomic force microscopes, and are generally smaller. This is often thought to be the result of non-conducting material on the sides of the pattern, following etch. From the perspective of lithography, there is also an etch bias between the resist and final dimensions which is relevant, and non-uniformities in the etch may introduce across-wafer variations, which need to be taken into account. Despite these complications, electrical linewidth measurements are very useful for characterizing linewidth control across lenses, because the speed of measurement enables a lot of data to be collected.

The structure in Fig. 8.11 can be used for measuring the widths of isolated lines. Additional structures can be added to provide measurements of linewidths of grouped structures, for any pitch that is desired. A method has even been developed for measuring contact sizes using electrical resistance measurements,[24] by measuring the effective electrical linewidth W of a structure with holes (Fig. 8.12), compared to the electrical linewidth W_{ref} of an identical reference structure without holes. The diameter d of the holes is given by:

$$d = \frac{W_{ref}}{\sqrt{12\pi}}\left[\sqrt{1 + \frac{48}{N}\left(\frac{L}{W} - \frac{L}{W_{ref}}\right)} - 1\right]^{\frac{1}{2}},\qquad(8.8)$$

where N is the number of contacts in the structure being measured.

Electrical metrology can also be used for monitoring SEM stability.[25] Even though electrical linewidth measurements do not provide absolute measurements, a correlation between electrical and SEM measurements provides a method for checking the stability of the SEMs, through comparison with electrical measurements.

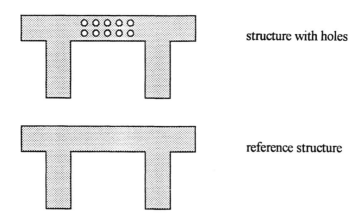

structure with holes

reference structure

Figure 8.12. Configuration for measuring contact diameter electrically.

8.4 MEASUREMENT ERROR BUDGETS

Following measurement there are four possible outcomes, indicated in Table 8.1. Alpha errors, in which good products are measured as bad products, result in unnecessary rework, while beta errors cause bad products to be accepted as good and will result in yield loss. Processes that are barely capable will have a high level of alpha errors, because of measurement imprecision. This is a direct cost of inadequate measurement capability, coupled with marginal process capability.[26] These costs will occur whenever process capability is low, and measurements are required to disposition material. From this perspective, good metrology can result in cost reductions.

Actual product quality ↓	Measured product quality	
	Good	Bad
Good	Correct decision	Alpha error
Bad	Beta error	Correct decision

Table 8.1. The types of outcomes from measurements of products.

The current level of repeatability for SEMs used for linewidth measurement is 5 nm (3σ).[27] The level of adequacy for this value needs to be determined. Consider metrology metrics used in other industries. For example, the Automotive Industry Action Group defines a measurement method with *good precision* as one in which:

$$5.15\sigma_{meas} < 0.3(USL - USL),\tag{8.9}$$

where σ_{meas} is the standard deviation of the measurement, and USL and LSL are the upper and lower specification limits for the process, respectively.[28] Similarly, a measurement method with *very good precision* is one with

$$5.15\sigma_{meas} < 0.1(USL - USL).\tag{8.10}$$

For a 250 nm lithography process requiring +/-10% linewidth control, current SEMs have good, but not very good, precision. Recent analysis has shown that measurement systems provide correct decisions according to[29]:

$$P_{correct} = 1 - \frac{\arctan\left(\dfrac{\sigma_{meas}}{\sigma_{process}}\right)}{\pi}\left[e^{-0.5\left(\frac{\mu - LSL}{\sigma_{process}}\right)^2} + e^{-0.5\left(\frac{USL - \mu}{\sigma_{process}}\right)^2}\right]\tag{8.11}$$

where $\sigma_{process}$ is the process standard deviation and μ is the process mean. For example, consider a 250 nm lithography process with $3\sigma_{process}$ = 23 nm linewidth control, requirements of +/- 25 nm (10%) linewidth control about the 250 nm target value, and a measurement system with 5 nm ($3\sigma_{meas}$) precision and no mean offset. In this example, the correct decision will be made 96% of the time. This means that 4% of processed material is either reworked or has degraded yield, because of the measurement imprecision. When processes are marginal, as often occurs in advanced semiconductor processing, measurement systems with very good precision (Eq. 8.10) are required to avoid high levels of unnecessary rework and degraded yield.

8.5 MEASUREMENT OF OVERLAY

Overlay is usually measured using optical linewidth measurement systems. Consider the structure shown in Fig. 8.13. The linewidths X_1 and X_2 can be used to calculate the overlay in the X direction:

$$\Delta X = \frac{1}{2}\left(X_2 - X_1\right) \tag{8.12}$$

The features shown in Fig. 8.13 are sufficiently large that optical linewidth measurement systems can be used, with their throughput advantage over SEMs. Sufficient data can be collected automatically in order to perform the overlay modeling discussed in Chapter 5.

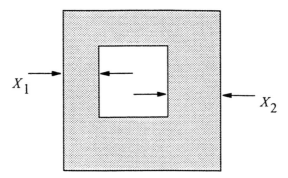

Figure 8.13. A structure for measuring overlay. The inner box is part of the substrate pattern, and the outer box is created in the overlaying photoresist layer.

Overlay *measurement* can introduce apparent overlay errors. The most common of these errors will be discussed in this section. This error is called tool-induced shift (TIS), and it results from asymmetries in the optics of the overlay measurement system. Because the materials of Layer 1 and Layer 2 in Fig. 8.13 are not the same, the measurements of X_1 and X_2 are not equivalent if there are asymmetries in the measurement equipment. Types of asymmetries that can occur are tilted or decentered lenses, nonuniform illumination, lens aberrations, and nonuniform detector response.[30] The presence of TIS is easily verified by rotating wafers from $0°$ to $180°$.[31,32] In the absence of TIS

$$\Delta X_{(0)} \rightarrow -\Delta X_{(180)}, \tag{8.13}$$

where $\Delta X_{(0)}$ is the measurement for the unrotated wafer and $\Delta X_{(180)}$ is the measurement for the wafer rotation $180°$. A measure of TIS is therefore

$$TIS = \Delta X_{(0)} + \Delta X_{(180)}, \tag{8.14}$$

which is ideally zero. The TIS value can be measured when overlay measurement programs are first established for particular process layers, and automatic compensation can be made to reduce the TIS error.[33] Asymmetries in alignment measurement can also be induced when features are too close to each other (X_1 or X_2 is too small), relative to the resolving power of the optics in the measurement tool.[34] The resolution of optical tools using visible light is typically in the range of $0.8 - 1.0$ µm, while systems using ultraviolet light may have somewhat lower resolution.

Wafer processing can induce asymmetries in overlay measurement marks. Consider situations in which metal is sputtered onto wafers. If the deposition is not collimated, metal may build up preferentially on one side of the overlay measurement marks near the edges of the wafers (Fig. 5.14). Measurements of overlay will be shifted relative to the overlay of the critical features. Errors in overlay measurement caused by wafer processing are referred to as wafer-induced shift (WIS). For the example illustrated in Fig. 5.14, the overlay error will be in opposite directions on opposite sides of the wafer and will appear as a wafer scaling error.[35] To the extent that asymmetries in step coverage are repeatable, wafer-induced-shifts can be calibrated by comparing overlay measurements before and after etch.[36]

Chemical-mechanical polishing (CMP), which causes problems for correct acquisition of alignment targets (Chapter 5), also creates difficulties for overlay measurement. Overlay measurement structures need to be optimized to minimize this effect. The bar or frame structures shown in Fig. 8.14 typically lead to less of a problem than the box-in-box structures, though the exact dimensions need to be determined for individual processes.

As discussed in Chapter 5, intrafield registration depends upon the illumination conditions and the feature type. Lines and spaces may have different registration at the different illumination settings. This will cause a problem for metrology if the critical features are one type, such as contact holes, while other types of features, such as lines, are used for overlay measurement.[37] This type of subtle difference will be significant as overlay requirements approach 50 nm or less.

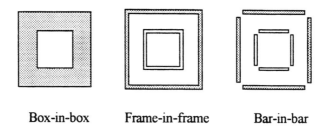

Box-in-box   Frame-in-frame   Bar-in-bar

Figure 8.14. Various structures for measuring overlay.

[1]   L. Galbraith and A. Neukermans, "Contamination Detection on Semiconductor Wafers," SPIE Vol. 774, pp. 13 – 20 (1987).

[2]   K. M. Monahan, M. Davidson, Z. Grycz, R. Krieger, B. Sheumaker, and R. Zmrzli, "Low-Loss Electron Imaging and its Application to Critical Dimension Metrology," SPIE Vol. 2196, pp. 138 – 144 (1994).

[3]   L. Reimer, *Image Formation in Low-voltage Scanning Electron Microscopy*, SPIE Press, Vol. TT12 (1993).

[4]   R. R. Hershey and M. B. Weller, "Nonlinearity in Scanning Electron Microscope Critical Dimension Measurements Introduced by the Edge Detection Algorithm," SPIE Vol. 1926, pp. 287 – 294 (1993).

[5]   J. I. Goldstein, D. E. Newbury, P. Echlin, D. C. Joy, C. Fiori, and E. Lifshin, *Scanning Electron Microscopy and X-Ray Microanalysis,* 2nd Edition, Plenum Press, New York (1984).

[6]   J. Finders, K. Ronse, L. Van den Hove, V. Van Driessche, and P. Tzviatkov, "Impact of SEM Accuracy on the CD-control During Gate Patterning Process of 0.25 μm Generations," Proceedings of the Olin Microlithography Seminar, pp. 17 – 30 (1997).

[7]   K. M. Monahan, J. P. H. Benschop, and T. A. Harris, "Charging Effects in Low-voltage SEM Metrology," SPIE Vol. 1464, pp. 2 – 9 (1991).

[8]   D. C. Joy and C. S. Joy, "Low Voltage Scanning Electron Microscopy," *Micron*, Vol. 27, No. 3 – 4, pp. 247 – 263 (1996).

[9] M. Davidson and N. T. Sullivan, "An Investigation of the Effects of Charging in SEM Based CD Metrology," SPIE Vol. 3050, pp. 226 – 242 (1997).

[10] C. M. Cork, P. Canestrari, P. DeNatale, and M. Vasconi, "Near and Sub-Half Micron Geometry SEM Metrology Requirements for Good Process Control," SPIE Vol. 2439, pp. 106 – 113 (1995).

[11] S. R. Rogers, "New CD-SEM Technology for 0.25 μm Production," SPIE Vol. 2439, pp. 353 – 362 (1995).

[12] NIST SRM standards are available from the Office of Standard Reference Materials, NIST, EM 205, Gaithersburg, MD 20899. Their phone number is (301) 975-6776. Information can be obtained from the NIST web site, www.nist.gov.

[13] M. T. Postek, A. E. Vladar, S. Jones, and W. J. Keery, "Report on the NIST Low Accelerating Voltage SEM Magnification Standard Interlaboratory Study," SPIE Vol. 1926, pp. 268 – 286 (1993).

[14] M. T. Postek, "Scanning Electron Microscope-based Metrological Electron Microscope System and New Prototype Scanning Electron Microscope Magnification Standard," *Scanning Microscopy*, 3(4), pp. 1087 – 1099 (1989).

[15] Y. Nakayama and K. Toyoda, "New Submicron Dimension Reference for Electron-Beam Metrology System," SPIE Vol. 2196, pp. 74 – 84 (1994).

[16] B. L. Newell, M. T. Postek, and J. P. van der Ziel, "Performance of the Protoype NIST SRM 2090A SEM Magnification Standard in a Low-Accelerating Voltage SEM," SPIE Vol. 2439, pp. 383 – 390 (1995).

[17] T. W. Reilly, "Metrology Algorithms for Machine Matching in Different CD SEM Configurations," SPIE Vol. 1673, pp. 48 – 55 (1992).

[18] K. Phan, J. Nistler, and B. Singh, "Metrology Issues Associated with Submicron Linewidths," SPIE Vol. 1464, pp. 424 – 437 (1991).

[19] E. E. Chain, M. G. Ridens, and J. P. Annand, "SPC Qualification Strategy for CD Metrology," SPIE Vol. 2876, pp. 218 – 224 (1996).

[20] D. A. Chernoff, "Atomic Force Microscope (AFM) Analysis of Photoresist Test Structures for use in SEM as In-House Linewidth Standards," SPIE Vol. 2439, pp. 392 – 400 (1995).

[21] M. T. Postek, A. E. Vladar, and M. P. Davidson, "Fourier Transform Feedback Tool for Scanning Electron Microscopes Used in Semiconductor Metrology," SPIE Vol. 3050, pp. 68 – 79 (1997).

[22] L. J. Zych, G. Spadini, T. F. Hassan, and B. A. Arden, "Electrical Methods for Precision Stepper Column Optimization," SPIE Vol. 633, pp. 98 – 105 (1986).

[23] L. W. Linholm, R. A. Allen, and M. W. Cresswell, "Microelectronic Test Structures for Feature Placement and Electrical Linewidth Metrology," in *Handbook of Critical Dimension Metrology and Process Control,* K. M. Monahan, ed., SPIE Press, Bellingham, WA (1993).

[24] B. J. Lin, J. A. Underhill, D. Sundling, and B. Peck, "Electrical Measurement of Submicrometer Contact Holes," SPIE Vol. 921, pp. 164 – 169 (1988).

[25] E. E. Chain and M. Griswold, "In-Line Electrical Probe for CD Metrology," SPIE Vol 2876, pp. 135 – 146 (1996).

[26] S. M. Kudva and R. W. Potter, "Cost Analysis and Risk Assessment for Metrology Applications," SPIE Vol. 1673, pp. 2 – 13 (1992).

[27] R. R. Hershey and R. C. Elliott, "Procedure for Evaluating Measurement System Performance: A Case Study," SPIE Vol. 2439, pp. 363 – 373 (1995).

[28] *Measurement Systems Analysis Reference Manual,* Automotive Industry Action Group, Detroit, MI (1990).

[29] J. Engel and B. de Vries, "Evaluating a Well-Known Criterion for Measurement Precision," *J. Quality Technol.,* 29(4), pp. 469 – 476 (1997).

[30] R. M. Silver, J. Potzick, and R. D. Larrabee, "Overlay Measurements and Standards," SPIE Vol. 3429, pp. 262 – 272 (1995).

[31] D. J. Coleman, P. J. Larson, A. D. Lopata, W. A. Muth, and A. Starikov, "On the Accuracy of Overlay Measurements: Tool and Mark Asymmetry Effects," SPIE Vol. 1261, pp. 139 – 161 (1990).

[32] A. Starikov, D. J. Coleman, P. J. Larson, A. D. Lopata, and W. A. Muth, "Accuracy of Overlay Measurements: Tool and Mark Asymmetry Effects," *Optical Engineering,* Vol. 31, pp. 1298 – 1310 (1992).

[33] M. E. Preil, B. Plambeck, Y. Uziel, H. Zhou, and M. W. Melvin, "Improving the Accuracy of Overlay Measurements through Reduction in Tool and Wafer Induced Shifts," SPIE Vol. 3050, pp. 123 – 134 (1997).

[34] N. Smith, G. Goelzer, M. Hanna, and P. Troccolo, "Minimizing Optical Overlay Measurement Errors," SPIE Vol. 1926, pp. 450 – 462 (1993).

[35] J-S. Han, H. Kim, J-L. Nam, M-S Han, S-K. Lim, S. D. Yanowitz, N. P. Smith, and A. M. C. Smout, "Effects of Illumination Wavelength on the Accuracy of Optical Overlay Metrology," SPIE Vol. 3051, pp. 417 – 425 (1997).

[36] Y. Tanaka, M. Kamiya, and N. Suzuki, "New Methodology of Optimizing Optical Overlay Measurement," SPIE Vol. 1926, pp. 429 – 439 (1993).

[37] T. Saito, H. Watanabe, and Y. Okuda, "Overlay Error of Fine Patterns by Lens Aberration using Modified Illumination," SPIE Vol. 3051, pp. 687 – 696 (1997).

CHAPTER 9

CONTROL OF OPERATIONS

Once a set of superior processes and control methodologies has been developed, additional actions are required for these to be made operational. Implementation ultimately involves a large number of operators, technicians, engineers, and supervisors, all of whom need to apply techniques consistently and must work in coordination with each other. It is a sad fact that excellent technology has often failed to become a reality in manufacturing because of inadequate implementation. Operational considerations, such as documentation, are essential for successfully controlling processes, and lithography engineers and managers must address these issues as well as purely technical matters.

Unsuccessful implementation often results even when attention is paid to operational matters, but inadequate methods are used. The first section of this chapter covers the theory of self-control, which provides a framework for engineers and managers to implement processes and procedures successfully. A key implementation tool, documentation, is covered in the next section. This chapter, and the text, is concluded with a discussion of ISO 9000, the quality standard which identifies the most essential elements of any quality control program.

9.1 SELF-CONTROL

It is essential to consider the human aspects of processes. The idea of controlling the actions of people has acquired a bad reputation during the past few decades, largely as a consequence of the implementation of specific practices in particular industries. Most notable were the American automobile manufacturers a few decades ago, who operated according to top-down management principles and suffered from management-worker conflicts, high manufacturing costs, and poor quality. The application of the scientific management principles of Frederick Taylor were often blamed for the poor competitive position of the American automobile industry. At the same time that these criticisms were being made, some of the most acclaimed and successful businesses were making use of strict work instructions, and well-defined procedures were found to be a key component of some of the best Japanese practices. For example, employee suggestions have been a key Japanese business practice, and it was observed by Kenjiro Yamada, the managing director of the Japan Human Relations Association, that the number of suggestions was higher in situations in which work instructions were rigidly fixed.[1] The control of operations is essential for the efficient manufacturing of quality products, but failure can result if this control is implemented incorrectly. This chapter will contain an overview of the critical elements for controlling operations.

A key factor for a successful operation, where processes are controlled, but the workers are nevertheless empowered to make necessary decisions, is the extent to which operators have been provided with the means for controlling operations by themselves. An excellent set of guidelines for establishing operator self-control has been formulated by Joseph Juran.[2] According to Juran, operator self-control exists when all people involved with a process have been provided with the means for:

1) Knowing what they are supposed to do.
2) Knowing what they actually are doing.
3) Regulating the process if 1) and 2) disagree.

Managers and engineers who expect operators and technicians to make decisions need to ensure that these three requirements are met, and further discussion of these criteria is worthwhile.

Knowing what they are supposed to do.

This consists of product and process standards, which should be written specifications. For clean room operations, documentation is rarely available in printed-paper formats, and "written specifications" is meant to refer to documentation in concrete text and/or graphics formats. It is also important to ensure that responsibilities are clearly defined. All too often specifications contain statements such as "Resist coater bowls should be cleaned weekly." With three or four shift operations, it is easy for everyone to assume that someone else is performing the specified task. Training is also essential for satisfying this element of self-control, to ensure that people know how to execute the tasks that they have been assigned.

Knowing what they actually are doing.

Closed feedback loops are required between inspection/measurement and processing operations. The pace of the cleanroom is swift, so data need to be formatted to allow operators to review the data quickly. Sufficient information needs to be provided to guide corrective action. It is also important to report only a digestible number of important issues; otherwise, it will be too overwhelming to be effective. Statistical process control is a particularly powerful feedback system, providing information about the state of the process, as well as the parameters of the system when it is in control.

Ability to regulate.

The process must be capable of producing a product to specifications if operators are expected to be held accountable for producing a good product. Corrective action is most effective in situations where the people have understandings of common failure modes, and the process is responsive to regulatory action in a predictable cause-and-effect relationship. Resources must be readily available to the worker who regulates the process. For example, in one wafer fab, operators were required to routinely wipe down equipment with isopropanol and water. Because this solution contained a flammable material, management required that bottles containing the liquid be kept, when not in use, in a cabinet designed for storing flammable materials. Unfortunately, only one such cabinet was installed within a large facility, and no one was assigned the task of ensuring that a supply of full bottles would be maintained inside the cabinet. It was not surprising that busy operators who worked in the parts of the wafer fab far from the cabinet frequently failed to wipe down their equipment. Making it a requirement to wipe the equipment was insufficient, because the means for performing the task was supported inadequately. Resources needed to be made available which would have allowed the task to be performed easily and quickly.

Whenever there is an operational failure, such as when an incorrect procedure is performed, it is useful to review the extent to which the lack of operator self-control played a role. This review should follow a checklist based on the three conditions that collectively constitute operator self-control. Very often inadequacies in documentation, training, or the process itself can be identified as factors contributing to the problem. Identifying and then correcting these factors can prevent future failures.

9.2 DOCUMENTATION

Written instructions are a powerful quality tool. One engineer, who transferred from one location where written specifications were well supported to another location where documentation was not employed rigorously described the problem he was having with implementing improvements: "I don't know what they are really doing (on the fab floor), so I don't know what there is to change. It's like trying to get a solid grip on oatmeal." As noted previously, the number of suggestions in Japanese companies was higher when the working instructions were rigidly specified. Documenting work instructions has several advantages:

1) They are a centralized source of information, available to operators, engineers, and managers.

2) They provide a communications link. In a well organized company there is a level of assurance that operators are made aware of changes to practices on a timely basis. Documented work instructions provide a well-defined method for implementing quality and productivity improvement.

3) Closure for statistical process control is provided. SPC provides the means for identifying a loss of process control, but does not provide operators with guidance on what to do when control is lost. This is an essential part of the "ability to regulate."

Part of the discipline of documentation is a requirement for clarity in work instructions, and the document control system should contain training for those who write work instructions. Documented work instructions should inform those who perform operations as to what should be done. The reasons and motivations for the particular instructions should not be included in the work instructions; otherwise the documents will become too large, and it will become difficult to find information. It is usually worthwhile to explain the basis for work instructions, but this information should be provided to operators in special off-line training classes.

It has often been asserted that well-defined work instructions are appropriate for manufacturing organizations, but a development operation represents a different situation. Examples show that documentation is important in development.[3] In one instance, a company had produced working circuits after working for two years to develop the newest generation of DRAMs. Unfortunately, processes were frequently being changed, but not documented, and the company was unable to determine what processes had actually been used to produce the good chips. It was almost another year before they were able to replicate their initial success. In a different company the development organization waited until a new process was ready to be transferred to manufacturing before documentation was generated. The writing process took several months, delaying the introduction of the new technology.

Part of the reluctance to make use of a potentially powerful quality tool stems from how difficult it is to use many document control systems. All too often approvals take too long to obtain, or engineers encounter unexpected questions concerning their new processes. These problems can be avoided by establishing approval systems that involve pre-defined requirements and minimum levels of approval. The requirements for making changes during process development should be considerably different from those used during volume manufacturing, allowing for greater flexibility where it is needed.

Management support is essential for documentation to be a useful quality tool. First, people within an organization need to believe that adherence to specifications is expected. Document control systems will quickly become irrelevant if management does not insist that process changes be implemented through the document control system. E-mail is a particularly insidious technology, because it has the appearance of control ("I wrote the instructions exactly"), but rarely are records maintained of process changes "implemented" by e-mail.

The difference between a well-controlled and poorly-controlled process is often just a detail. Documented work instructions provide the means for ensuring that the details are correctly addressed. As mentioned earlier, engaging operators and technicians into quality improvement programs is facilitated by providing a concrete system for implementing changes. Documentation is a key element of any program to improve process control.

9.3 ISO 9000

ISO 9000 generally refers to a set of standards to which quality systems are expected to perform. These are not technical standards, which are necessarily specific to individual industries and products, but standards which apply to the management of organizations. The most comprehensive of these standards, ISO 9001, which is applicable to organizations which design, develop, manufacture, install, and service products, will be discussed here (see Table 9.1). Some companies or plant sites may be registered to ISO 9002, which is a subset of the ISO 9001 standard and is intended for organizations that only manufacture and install products. It is possible for one site of a company, where development is done, to be registered to the ISO 9001 standard, and another site, which does only manufacturing, to be registered to ISO 9002. A comprehensive discussion of the ISO 9000 standards would require an entire book, not just a chapter. It is the purpose of this section to highlight those elements of ISO 9001 of greatest significance for lithographers.

```
                  Elements of ISO 9001

        1)  Management Responsibility
        2)  Quality System
        3)  Contract Review
        4)  Design Control
        5)  Document and Data Control
        6)  Purchasing
        7)  Control of Customer-Supplied
            Product
        8)  Product Identification and
            Traceability
        9)  Process Control
        10) Inspection and Testing
        11) Control of Inspection,
            Measuring, and Test Equipment
        12) Inspection and Test Status
        13) Control of Nonconforming
            Product
        14) Corrective and Preventive Action
        15) Handling, Storage, Packaging,
            Preservation, and Delivery
        16) Control of Quality Records
        17) Internal Quality Audits
        18) Training
        19) Servicing
        20) Statistical Techniques
```

Table 9.1. The twenty elements, or sections, of ISO 9001.

The need to reduce redundant quality audits, in which customers tour facilities and peruse quality records, motivated the creation of the ISO 9000 standards. Such audits could occur several times in a single week.[4] Because there are many characteristics common to all quality systems, there is considerable redundancy among such audits. Under ISO 9000, companies conform to certain quality standards and are audited by accredited auditors. A registry of companies conforming to ISO 9000 addresses the problem of audit redundancy.

There are few requirements to conform to the ISO 9000 standards, outside of a handful of classes of regulated products in the European Union, relating primarily to safety concerns.[5] However, a standard for quality systems that has wide applicability must contain only the most essential elements for managing quality. Few customers want to buy from suppliers who are unable to demonstrate that they can meet these most basic of standards. For this reason, ISO 9000 has become a de facto necessity of doing business.

The basic elements of ISO 9001 are shown in Table 9.1. Only a few of these, of specific relevance to lithography, will be discussed in this text. Part of the Process Control section of the ISO 9001 standard reads:

4.9 PROCESS CONTROL
The supplier shall identify and plan the production, installation, and servicing processes which directly affect quality and shall ensure that these processes are carried out under controlled conditions. Controlled conditions shall include the following:
a) documented procedures defining the manner of production, installation, and servicing, where the absence of such procedures could adversely affect quality;
b) use of suitable production, installation, and servicing equipment, and a suitable working environment;
c) compliance with reference standards/codes, quality plans, and/or documented procedures;
d) monitoring and control of suitable process parameters and product characteristics;
e) approval of processes and equipment, as appropriate;
f) criteria for workmanship, which shall be stipulated in the clearest practical manner (e.g., written standards, representative samples, or illustrations);
g) suitable maintenance of equipment to ensure continuing process capability.

Documented work instructions are mentioned explicitly in this element of ISO 9001. Further details of the document and records control systems are contained in elements 5 and 16. Section 4.9 of ISO 9001 also calls out explicitly for process monitoring. Statistical process control and the methods described in Chapters 3 and 7 of this text provide the means for satisfying this requirement of ISO 9001.

The need for suitable measurement capability is stated explicitly in element 4.11, Control of Inspection, Measuring, and Test Equipment:

4.11.2 Control procedure
The supplier shall:
a) determine the measurements to be made and the accuracy required, and select the appropriate inspection, measuring, and test equipment that is capable of necessary accuracy and precision;
b) identify all inspection, measuring, and test equipment that can affect product quality, and calibrate and adjust them at prescribed intervals, or prior to use, against certified equipment having a known valid relationship to internationally or nationally recognized standards. Where no such standards exist, the basis used for calibration shall be documented;
c) define the process employed for the calibration of inspection, measuring, and test equipment...

This section clearly addresses the standards dilemma in critical dimension metrology. In order to meet the requirements of ISO 9000, metrologists must implement a suitable program for controlling the metrology tools. Further guidance for conforming to the metrology requirements of ISO 9000 can be found in another standard, ISO 10012, "Quality assurance requirements for measuring equipment."

ISO 9000 does explicitly mention the need for the use of appropriate statistical techniques:

4.20 STATISTICAL TECHNIQUES

4.20.1 Identification of need
The supplier shall identify the need for statistical techniques required for establishing, controlling, and verifying process capability and product characteristics.
4.20.2 Procedures
The supplier shall establish and maintain documented procedures to implement and control the application of the statistical techniques identified in 4.20.1.

Because it is intended that ISO 9000 have wide applicability to numerous industries, it is left to the technical experts within each industry to determine where statistical methods, such as statistical process control, should be implemented. This element of ISO 9000 brings us full circle to the preceding subjects covered in this text: the identification of processes where controls need to be inserted and the technical and statistical tools that can facilitate implementation. ISO 9000 recognizes that the expertise of engineers and technicians forms the foundation of quality control. This is a most certain truth for microlithography.

---

[1] M. Imai, *Kaizen: The Key to Japan's Competitive Success*, Random House, New York, p. 113 (1986).

[2] *Juran's Quality Control Handbook*, 4th Edition, J. M Juran and F. M. Gryna, eds., McGraw Hill, New York (1988).

[3] H. J. Levinson and J. Ben-Jacob, "Managing Quality Improvement on a Development Pilot Line," *Quality Management Journal*, Vol. 3(2), pp. 16 – 35 (1996).

[4] C. A. De Angelis, "ICI Advanced Materials Implements ISO 9000 Program," *Quality Progress*, Vol. 24, pp. 49 – 51 (1991).

[5] H. J. Levinson, "ISO 9000: What Every Microlithographer Should Know," SPIE Vol. 2196, pp. 536 – 550 (1993).

GENERAL REFERENCES

GENERAL REFERENCES ON QUALITY CONTROL:

- W. Shewhart, *Economic Control of Quality of Manufactured Product*, D. van Nostrand Co., New York, (1931).

- Edwards Deming, *Out of the Crisis*, MIT Center for Advanced Engineering Study, Cambridge, MA (1982).

- D. C. Montgomery, *Introduction to Statistical Quality Control*, John Wiley & Sons, New York (1996).

- Western Electric's *Statistical Quality Control Handbook*, Delmar Printing Company, Charlotte (1985).

- A. J. Duncan, *Quality Control and Industrial Statistics*, 5th Edition, Irwin, Homewood, Illinois (1986).

- M. G. Natrella, *Experimental Statistics*, U.S. Government Printing Office, Washington, D. C. (1966).

- S. S. Shapiro, *How to Test Normality and Other Distributional Assumptions*, American Society for Quality Control, Milwaukee (1986).

- G. E. P. Box, G. M. Jenkins, and G. C. Reinsel, *Time Series Analysis*, 3rd Edition, Prentice Hall, Englewood Cliff, NJ (1994).

- *Guide for Reducing Quality Costs*, 2nd Edition, ASQ Press, Milwaukee (1987).

- V. Barnett and T. Lewis, *Outliers in Statistical Data*, 3rd Edition, John Wiley and Sons, Chichester (1995).

- M. Imai, *Kaizen: The Key to Japan's Competitive Success*, Random House, New York, p. 113 (1986).

- *Juran's Quality Control Handbook*, 4th Edition, J. M Juran and F. M. Gryna, eds., McGraw Hill, New York (1988).

- K. Ishikawa, *Guide to Quality Control*, Asian Productivity Organization, White Plains, New York (1991).

- G. E. P. Box, W. G. Hunter, and J. S. Hunter, *Statistics for Experimenters*, John Wiley & Sons, New York (1978).

- G. E. P. Box and N. R. Draper, *Empirical Model-building and Response Surfaces*, John Wiley & Sons, New York (1987).

- G. Taguchi, *System of Experimental Design, Vols. 1 and 2*, American Supplier Institute Press, Dearborn, Michigan (1987).

PROCESS CAPABILITY:

- V. E. Kane, "Process Capability Indices," *Journal of Quality Technology*, Vol. 18, pp. 41 – 52 (1986).

- L. K. Chan, S. W. Cheng, and F. A. Spiring, "A New Measure of Process Capability: C_{pm}," *Journal of Quality Technology*, Vol. 20, pp. 162 – 175 (1988).

- R. A. Boyles, "The Taguchi Capability Index," *Journal of Quality Technology*, Vol. 23, pp. 17 – 26 (1991).

- D. J. Wheeler, *A Japanese Control Chart*, SPC Press, Knoxville (1986).

- R. McFadden, "Six-Sigma Quality Programs," *Quality Progress*, pp. 37 – 42 (June, 1993).

- Y-M. Chou, D. B. Owen, and S. A. Borrego, "Lower Confidence Limits on Process Capability Indices," *Journal of Quality Technology*, pp. 223 – 229 (1990).

- F. S. Hillier, "$\overline{X}$ - and R-Chart Control Limits Based on a Small Number of Subgroups," *Journal of Quality Technology*, 1, pp. 17 – 26 (1969).

- C. P. Quesenberry, "SPC Q Carts for Start-Up Processes and Short or Long Runs," *Journal of Quality Technology*, 23(3), pp. 213 – 224 (1991).

- P. P. Ramsey and P. H. Ramsey, "Simple Tests of Normality in Small Samples," *Journal of Quality Technology*, pp. 299 – 313 (1990).

- J. F. MacGregor, "A Different View of the Funnel Experiment," *Journal of Quality Technology*, Vol. 22, pp. 255 – 259 (1990).

- S. T. Mandraccia, G. D. Halverson, and Y-M. Chou, "Control Chart Design Strategies for Skewed Data," SPIE Vol. 2876, pp. 196 – 205 (1996).

LINEWIDTH CONTROL:

- D. Heberling, "Litho Equipment Matching with E_0," Proceedings of the KTI Microelectronics Seminar, pp. 233 – 243 (1990).

- C. Yu, T. Maung, C. Spanos, D. Boning, J. Cheung, H-Y Liu, K-J. Chang, and D. Bartelink, "Use of Short-Loop Electrical Measurements for Yield Improvement," *IEEE Trans. Semicond. Manuf.*, Vol. 8(2), pp. 150 – 159 (1995).

- R. C. Elliott, R. R. Hershey, and K. G. Kemp, "Cycle-time Reduction of CD Targeting using Automatic Metrology and Analysis," SPIE Vol. 2439, pp. 70 – 77 (1995).

- C. Takemoto, D. Ziger, W. Connor, and R. Distasio, "Resist Tracking: A Lithographic Diagnostic Tool," SPIE Vol. 1464, pp. 206 – 214 (1991).

- M. van den Brink, C. G. M. de Mol, H. F. D. Linders, and S. Wittekoek, "Matching Management of Multiple Wafer Steppers Using a Stable Standard and a Matching Simulator," SPIE Vol. 1087, pp. 218 – 232 (1989).

- K. Kemp, C. King, W. Wu, and C. Stager, "A "Golden Standard" Wafer Design for Optical Stepper Characterization," SPIE Vol. 1464, pp. 260 – 277 (1991).

- H. J. Levinson and W. H. Arnold, "Focus: The Critical Parameter for Sub-micron Lithography," *J. Vac. Sci. Technol.*, B5(1), pp. 293 – 298 (1987).

- K. Ronse, R. Pforr, L. van den Hove, and M. Op de Beeck, "CD Control: the Limiting Factor for i-line and Deep-UV Lithography?" OCG Microelectronics Seminar, pp. 241 – 254 (1995).

- K. Ronse, M. Op de Beeck, A. Yen, K-H. Kim, and L. van den Hove, "Characterization and Optimization of CD Control for 0.25 mm CMOS Applications," SPIE Vol. 2726, pp. 555 – 563 (1996).

- P. Schoenborn and N. F. Pasch, "Process Sensitivity Analysis: Applications to Photolithography," SPIE Vol. 1087, pp. 290 – 298 (1989).

- Z. Krivokapic, W. D. Heavlin, and D. Kyser, "Process Capabilities of Critical Dimensions at Gate Mask," SPIE Vol. 2440, pp. 480 – 491 (1995).

- R. W. Leonhardt and T. R. Scott, "Deep-UV Excimer Laser Measurements at NIST," SPIE Vol. 2439, pp. 448 – 459 (1995).

- K. H. Kim, W. S. Han, C. H. Kim, H. Y. Kang, C. G. Park, and Y. B. Koh, "Characteristics of Standing Wave Effect of Off-axis Illumination Depending on two Different Resist Systems and the Polarization Effect of Stepper," SPIE Vol. 2197, pp. 42 – 53 (1994).

- J. Sturtevant and B. Roman, "Antireflection Strategies for Advanced Photolithography," *Microlithography World*, pp. 13 – 21 (1995).

- G. MacBeth, "Thermal Effects in Photoresist Coating Processes," KTI Microelectronics Seminar, pp. 327 – 340 (1988).

- S. Dick and B. Greenstein, "Improved Photolithography Process Performance Through the Use of an Integrated Photosector," KTI Microelectronics Seminar, pp. 1 – 8 (1989).

- M. Reihani, "Environmental Effects on Resist Thickness Uniformity," Semiconductor International, pp. 120 – 121 (June, 1992).

- J. S. Petersen and J. D. Byers, "Examination of Isolated and Grouped Feature Bias in Positive Acting Chemically Amplified Resist Systems," SPIE Vol. 2724, pp. 163 – 171 (1996).

- J. S. Petersen, C. A. Mack, J. W. Thackeray, R. Sinta, T. H. Fedynyshyn, J. J. Mori, J. D. Byers, and D.A. Miller, "Characterization and Modeling of a Positive Acting Chemically Amplified Resist," SPIE Vol. 2438, pp. 153 – 166 (1995).

- J. M. Kulp, "CD Shift Resulting from Handling Time Variation in the Track Coat Process," SPIE Vol. 1466, pp. 630 – 640 (1991).

- G. MacBeth, "Prebaking Positive Photoresists," Proceedings of the Kodak Microelectronics Seminar, pp. 87 – 92 (1982).

- O. D. Crisalle, C. L. Bickerstaff, D. E. Seborg, and D. A. Mellichamp, "Improvements in Photolithography Performance by Controlled Baking," SPIE Vol. 921, pp. 317 – 325 (1988).

- C. M. Garza, C. R. Szmanda, and R. L Fischer, "Resist Dissolution Kinetics and Submicron Process Control," SPIE Vol. 920, pp 321 – 338 (1988).

- M. K. Templeton, J. B. Wickman, and R. L. Fischer, Jr., "Submicron Resolution Automated Track Development Processes, Part 1: Static Puddle Development," SPIE Vol. 921 pp. 360 – 372 (1988).

- E. Bokelberg and W. Venet, "Effects of Relative-humidity Variation on Photoresist Processing," SPIE Vol. 2438, pp. 747 – 752 (1995).

- W. H. Arnold and H. J. Levinson, "High Resolution Optical Lithography Using an Optimized Single Layer Photoresist Process," Proceedings of the Kodak Microelectronics Seminar, pp. 80 – 92 (1983).

- J. A. Bruce, S. R. DuPuis, and H. Linde, "Effect of Humidity on Photoresist Performance," Proceedings of the OCG Microlithography Seminar, pp. 25 – 41 (1995).

- S. A. MacDonald, C. G. Wilson, and J. M. J. Frêchet, "Chemical Amplification in High-Resolution Imaging Systems," *Accounts of Chemical Research*, Vol. 27(6), pp. 151 – 158 (1994).

- K. R. Dean, D. A. Miller, R. A. Carpio, and J. S. Petersen, "Airborne Contamination of DUV Photoresists: Determining the New Limits of Processing Capability," Proceedings of the Olin Microlithography Seminar, pp. 109 – 125 (1996).

- K. van Ingen Schenau, M. Reuhman, and S. Slonaker, "Investigation of DUV Process Variables Impacting Sub-Quarter Micron Imaging," Proceedings of the Olin Microlithography Seminar, pp. 63 – 80 (1996).

- Y-K. Hsiao, C-H. Lee, S-L. Pan, K-L. Lu, and J-C. Yang, "A Study of the Relation Between Photoresist Thermal Property and Wettability," Proceedings of the Olin Microlithography Seminar, pp. 315 – 323 (1996).

- H. Ito, W. P. England, H. J. Clecak, G. Breta, H. Lee, D. Y. Yoon, R. Sooriyakumaran, and W. D. Hinsberg, "Molecular Design for Stabilization of Chemical Amplification Resist Toward Airborne Contamination," SPIE Vol. 1925, pp. 65 – 75 (1993).

- W. Maurer, K. Satoh, D. Samuels, and T. Fischer, "Pattern Transfer at $k_1 = 0.5$: Get 0.25 μm Lithography Ready for Manufacturing," SPIE Vol. 2726, pp. 113 – 124 (1996).

- K. Petrillo, "Process Optimization of Apex-E," SPIE Vol. 1926, pp. 176 – 187 (1993).

- S. G. Hansen, G. Dao, H. Gaw, Q-D. Qian, P. Spragg, and R. J. Hurditch, "Study of the Relationship Between Exposure Margin and Photolithographic Process Latitude and Mask Linearity," SPIE Vol. 1463, pp. 230 – 244 (1991).

RESIST THICKNESS CONTROL:

- D. P. Birnie III, B. J. J. Zelinski, S. P. Marvel, S. M. Melpolder, and R. L. Roncone, "Film/Substrate/Vacuum-Chuck Interactions During Spin-Coating," *Optical Engineering*, Vol. 31(9), pp. 2012 – 2020 (1992).

- R. Murray, P. T. Edwin, V. Taburaza, and J. Olin, "Airflow Controller Improves Photoresist Spin/Coat Uniformity," *Semiconductor International*, pp. 172 –173 (April, 1987).

- L. Matter, J. Zook, M. Hinz, J. Banas, and S. Ibrani, "New Coat Bowl Design Improves Photoresist Uniformity and Decreases Particle Contamination," Olin Microlithography Seminar Poster Session (1997).

- X. Zhu, F. Liang, A. Haji-Sheikh, and N. Ghariban, "A Computational and Experimental Study of Spin Coater Air Flow," SPIE Vol. 3333, pp. 1441 – 1451 (1998).

- W. J. Daughton, P. O'Hagan, and F. L. Givens, "Thickness Variance of Spun-On Photoresist, Revisited," Proceedings of the Kodak Microelectronics Seminar, pp. 15 – 20 (1978).

- "Lithography Track Thermal Metrology Update," *Semiconductor International* (1998).

- D. Boutin, A. Blash, J. P. Caire, D. Poncet, P. Fanton, M. Danielou, and B. Previtali, "Resist Coating Optimization on Eight Inches Deep UV Litho

Cell Modelization and Application to 0.25 μm Technology," SPIE Vol. 2439, pp. 495 – 502 (1995).

FEEDBACK AND AUTOMATIC PROCESS CONTROL:

- H. J. Levinson, "Control and Improvement of Complex Processes," *Quality Engineering*, Vol. 5(1), pp. 93 – 106 (1992).

- T. Batchelder, M. Ha, R. Haney, and W. Lee, "Sub-micron Linewidth Control with Automatic Optical Monitors," KTI Microelectronics Seminar, pp. 231 – 242 (1989).

- T. Batchelder, M. Ha, R. Haney, and W. Lee, "Sub-Micron Linewidth Control with Automatic Optical Monitors," KTI Microelectronics Seminar, pp. 231 – 242 (1989).

- M. Watts and S. Williams, "A Novel Method for the Prediction of Process Sensitivity in Photolithography," SPIE Vol. 1261, pp. 345 – 359 (1990).

- K. Kemp, D. Williams, J. Daggett, J. Cayton, S. Slonaker, and R. Elliott, "Critical Dimension Performance Characterization of an Advanced DUV Process Cell," Olin Microlithography Seminar, pp. 99 – 108 (1996).

- M. Drew and K. Kemp, "Automatic Feedback Control to Optimize Stepper Overlay," SPIE Vol. 1926, pp. 422 – 428 (1993).

- C. P. Ausschnitt, A. C. Thomas and T. J. Wiltshire, "Advanced DUV Photolithography in a Pilot Line Environment," *IBM J. Res. Develop.* Vol. 41(1/2), pp. 21 – 36 (1997).

- F. E. Grubbs, "An Optimum Procedure for Setting Machines or Adjusting Processes," *Journal of Quality Technology*, Vol. 15(4), pp. 186 – 189 (1983).

- S. V. Crowder, "Design of Exponentially Weighted Moving Average Schemes," *Journal of Quality Technology*, Vol. 21, pp. 155 – 162 (1989).

- J. S. Hunter, "The Exponentially Weighted Moving Average," *Journal of Quality Technology*, Vol. 18, pp. 203 – 210 (1986).

- J. M. Lucas and M. S. Saccucci, "Exponentially Weighted Moving Average Control Schemes: Properties and Enhancements," *Technometrics*, Vol. 32, pp. 1 – 12 (1990).

- K. L. Konnerth and F. H. Dill, "In-situ Measurement of Dielectric Thickness During Etching or Developing Processes," *IEEE Trans. Electron Devices*, Vol. ED-22, pp. 452 – 456 (1975).

- L. J. Lauchlan, K. M. Sautter, and T. Batchelder, "Automatic Process Control for VLSI Linewidth," *Solid State Technology*, pp. 333 – 337 (April 1985).

- L. J. Lauchlan, K. Sautter, T. Batchelder, and J. Irwin, "In-Line Automatic Photoresist Process Control," SPIE Vol. 539, pp. 227 – 233 (1985).

- C. Nygren, J. Daggett, and J. Grambow, "The Use of Develop End Point Detection to Eliminate Photolithography Process Variation," Proceedings of the KTI Microlithography Seminar, pp. 315 – 348 (1990).

- P. D. Flanner III, "Improved Methods of Photoresist Development Characterization," Proceedings of the KTI Microelectronics Seminar, pp. 231 – 238 (1987).

- K. M. Sautter, M. Ha, and T. Batchelder, "Development Process Control and Optimiziation Utilizing anEnd Point Monitor," Proceedings of the KTI Microelectronics Seminar, pp. 99 – 112 (1988).

- J. A. Bruce and B. J. Lin, "Determination of Exposure Dose by Photoresist Development Rate," Proceedings of the KTI Microelectronics Seminar, pp. 1 – 11 (1987).

- M. L. Miller and D. A. Mellichamp, "Development of an End-Point Detection Procedure for the Post-Expose Bake Process," SPIE Vol. 2439, pp. 78 – 88 (1995).

- S. Zaidi, S. L. Prins, J. R. McNeil, and S. S. H. Naqvi, "Metrology Sensors for Advanced Resists," SPIE Vol. 2196, pp. 341 – 351 (1994).

- J. L. Sturtevant, S. J. Holmes, T. G. VanKessel, P. D. Hobbs, J. C. Shaw, and R. R. Jackson, "Postexposure Bake as a Process-Control Parameter for Chemically Amplified Photoresist," SPIE Vol. 1926, pp. 106 – 114 (1993).

- J. L. Sturtevant, S. Holmes, T. G. Van Kessel, M. Miller, and D. Mellichamp, "Use of Scatterometric Latent Image Detector in Closed Loop Feedback Control of Linewidth," SPIE Vol. 2196, pp. 352 – 359 (1994).

- K. C. Hickman, S. M. Gaspar, K. P. Bishop, S. S. H. Haqvi, J. R. McNeil, G. D. Tipton, B. R. Stallard, and B. L. Draper, "Use of Diffracted Light from Latent Images to Improve Lithography Control," SPIE Vol. 1464, pp. 245 – 251 (1991).

- T. E. Adams, "Applications of Latent Image Metrology in Microlithography," SPIE Vol. 1464, pp. 294 – 312 (1991).

- S. L. Prins and J. R. McNeil, "Scatterometric Sensor for PEB Process Control," SPIE Vol. 2725, pp. 710 – 719 (1996).

- L. M. Milner, K. C. Hickman, S. M. Gaspar, K. P. Bishop, S. S. H. Haqvi, J. R. McNeil, M. Blain, and L. Draper, "Latent Image Exposure Monitor Using Scatterometry," SPIE Vol. 1673, pp. 274 – 283 (1992).

- M. L. Miller and D. A. Mellichamp, "Development of an End-Point Detection Procedure for the Post-Exposure Bake Process," SPIE Vol. 2439, pp. 78 – 88 (1995).

- T. Koizumi, T. Matsuo, M. Endo, and M. Sasago, "CD Control using Latent Image for Lithography," SPIE Vol. 2439, pp. 418 – 426 (1995).

- D. C. Montgomery, J. B. Keats, G. C. Runger, and W. S. Messina, "Integrating Statistical Process Control and Engineering Process Control," *Journal of Quality Technology*, Vol. 26(2), pp. 79 – 87 (1994).

- The April, 1997, Vol. 29(2) of the *Journal of Quality Technology* has a discussion on statistically-based process monitoring and control, with contributions from several authors and extensive references.

- G. Box and A. Luceño, *Statistical Control by Monitoring and Feedback Adjustment*, John Wiley & Sons, New York (1997).

FOCUS:

- T. O. Herndon, C. E. Woodward, K. H. Konkle, and J. I. Raffel, "Photocomposition and DSW Autofocus Correction for Wafer-Scale Lithography," Proceedings of the Kodak Microelectronics Seminar, pp. 118 – 123 (1983).

- A. Suzuki, S. Yabu, and M. Ookubo, "Intelligent Optical System for a New Stepper,"SPIE Vol. 772, pp. 58 – 65 (1987).

- J. E. van den Werf, "Optical Focus and Level Sensor for Wafer Steppers," *J. Vac. Sci. Technol.*, Vol. 10(2) , pp. 735 – 740 (1992).

- M. A. van den Brink, J. M. D. Stoeldraijer, and H. F. D. Linders, "Overlay and Field by Field Leveling in Wafer Steppers Using an Advanced Metrology System," SPIE Vol. 1673, pp. 330 – 344 (1992).

- J. W. Gemminck, "Simple and Calibratable Method for Determining Optimal Focus," SPIE Vol. 1088, pp. 220 – 230 (1989).

- S. Venkataram, C. Olejnik, G. Flores, and D. Tien, "An Automated Technique for Optimizing Stepper Focus Control," SPIE Vol. 2725, pp. 765 – 778 (1996).

- T. A. Brunner, A. L. Martin, R. M. Martino, C. P. Ausschnitt, T. H. Newman, and M. S. Hibbs, "Quantitative Stepper Metrology Using the Focus Monitor Test Mask," SPIE Vol. 2197, pp. 542 – 549 (1994).

- R. D. Mih, A. Martin, T. Brunner, D. Long, and D. Brown, "Using the Focus Monitor Test Mask to Characterize Lithographic Performance," SPIE Vol. 2440, pp. 657 – 666 (1995).

- E. R. Sherman and C. Harker, "Characterization and Monitoring of Variable NA and Variable Coherence Capable Photo Steppers Utilizing the Phase Shift Focus Monitor Reticle," SPIE Vol. 2439, pp. 61 – 69 (1995).

- T. A. Brunner and R. D. Mih, "Simulations and Experiments with the Phase Shift Focus Monitor," SPIE Vol. 2726, pp. 236 – 243 (1996).

- T. A. Brunner and S. M. Stuber, "Characterization and Setup Techniques for a 5X Stepper," SPIE Vol. 633, pp. 106 – 112 (1986).

- M. van den Brink, H. Franken, S. Wittekoek,and T. Fahner, "Automatic On-Line Wafer Stepper Calibration System," SPIE Vol. 1261, pp. 298 – 314 (1990).

- T. A. Brunner, S. Cheng, and A. E. Norton, "Stepper Image Monitor for Precise Setup and Characterization," SPIE Vol. 922, pp. 366 – 375 (1988).

- R. Pforr, S. Wittekoek, R. van den Bosch, L. van den Hove, R. Jonckheere, T. Fahner, and R. Seltmann, "In-process Image Detecting Technique for Determination of Overlay and Image Quality for the ASM-L Wafer Stepper," SPIE Vol. 1674, pp. 594 – 609 (1992).

- P. Dirksen, W. de Laat, and H. Megens, "Latent Image Metrology for Production Wafer Steppers," SPIE Vol. 2440, pp. 701 – 711 (1995).

- T. A. Brunner, J. M. Lewis, and M. P. Manny, "Stepper Self-Metrology Using Automated Techniques," SPIE Vol. 1261, pp. 286 – 297 (1990).

- K. Hale and P. Luehrman, "Consistent Image Quality in a High Performance Stepper Environment," Proceedings of the Kodak Microelectronics Seminar, pp. 29 – 46 (1986).

OVERLAY:

- R. Groves, "Statistics of Pattern Placement Errors in Lithography," *J. Vac. Sci Technol.*, B9(6), pp. 3555 – 3561 (1991).

- N. Magome and H. Kawai, "Total Overlay Analysis for Designing Future Aligner," SPIE Vol. 2440, pp. 902 – 912 (1995).

- M. A. van den Brink, J. M. D. Stoeldraijer, and H. F. D. Linders, "Overlay and Field by Field Leveling in Wafer Steppers Using an Advanced Metrology System," SPIE Vol. 1673, pp. 330 – 344 (1992).

- T. Saito, S. Sakamoto, K. Okuma, H. Fukumoto, and Y. Okuda, "Mask Overlay Scaling Error Caused by Exposure Energy Using a Stepper," SPIE Vol. 1926, pp. 440 – 449 (1993).

- J. D. Armitage, Jr. and J. P. Kirk, "Analysis of Overlay Distortion Patterns," SPIE Vol. 921, pp. 207 – 222 (1988).

- T. E. Zavecz, "Lithographic Overlay Measurement Precision and Calibration and Their Effect on Pattern Registration Optimization," SPIE Vol. 1673, pp. 191 – 202 (1992).

- V. Nagaswami and W. Geerts, "Overlay Control in Submicron Environment," Proceedings of the KTI Microelectronics Seminar, pp. 89 – 106 (1989).

- M. A. van den Brink, C. G. M. de Mol, and R. A. George, "Matching Performance for Multiple Wafer Steppers Using an Advanced Metrology Procedure," SPIE Vol. 921, pp. 180 – 197 (1988).

- J. G. Maltabes, M. C. Hakey, and A. L. Levine, "Cost/Benefit Analysis of Mix-and-Match Lithography for Production of Half-Micron Devices," SPIE Vol. 1927, pp. 814 – 826 (1993).

- R. K. Brimacombe, T. J. McKee, E. D. Mortimer, B. Norris, J. Reid, and T. A. Znotins, "Performance Characteristics of a Narrow Band Industrial Excimer Laser," SPIE Vol. 1088, pp. 416 – 422 (1989).

- H. J. Levinson and R. Rice, "Overlay Tolerances for VLSI Using Wafer Steppers," SPIE Vol. 922, pp. 82 – 93 (1988).

- M. A. van den Brink, C. G. M. de Mol, and J. M. D. Stoeldraijer, "Matching of Multiple Wafer Steppers for 0.35 mm Lithography, Using Advanced Optimization Schemes," SPIE Vol. 1926, pp. 188 – 207 (1993).

- J. C. Pelligrini, "Comparisons of Six Different Intrafield Control Paradigms in an Advanced Mix-and-Match Environment," SPIE Vol. 3050, pp. 398 – 406 (1997).

- H. J. Levinson, M. E. Preil, and P. J. Lord, "Minimization of Total Overlay Errors on Product Wafers Using an Advanced Optimization Scheme," SPIE Vol. 3051, pp. 362 – 373 (1997).

- N. R. Farrar, "Effect of Off-axis Illumination on Stepper Overlay," SPIE Vol. 2439, pp. 273 – 280 (1995).

- T. Saito, H. Watanabe, and Y. Okuda, "Effect of Variable Sigma Aperture on Lens Distortion and Its Pattern Size Dependence," SPIE Vol. 2725, pp. 414 – 423 (1996).

- A. M. Davis, T. Dooly, and J. R. Johnson, "Impact of Level Specific Illumination Conditions on Overlay," Proceedings of the Olin Microlithography Seminar, pp. 1 – 16 (1997).

- C. S. Lee, J. S. Kim, I. B. Hur, Y. M. Ham, S. H. Choi, Y. S. Seo, and S. M. Ashkenaz, "Overlay and Lens Distortion in a Modified Illumination Stepper," SPIE Vol. 2197, pp. 2 – 8 (1994).

- C-S. Lim, K-S. Kwon, D. Yim, D-H. Son, H-S. Kim, and K-H. Baik, "Analysis of Nonlinear Overlay Errors by Aperture Mixing Related to Pattern Asymmetry," SPIE Vol. 3051, pp. 106 – 115 (1997).

- M. E. Preil, T. Manchester, and A. Minvielle, "Minimization of Total Overlay Errors when Matching Non-concentric Exposure Fields," SPIE Vol. 2197, pp. 753 – 769 (1994).

- W. W. Flack, G. E. Flores, J. C. Pellegrini, and M. Merrill, "An Optimized Registration Model for 2:1 Stepper Field Matching," SPIE Vol. 2197, pp. 733 – 752 (1994).

- T. A. Brunner, "Impact of Lens Aberrations on Optical Lithography," *IBM J. Res. Develop.*, Vol. 41(1/2), pp. 57 – 67 (1997).

- R. Rogoff, S. S. Hong, D. Schramm, and G. Espin, "Reticle Specific Compensations to Meet Production Overlay Requirements for 64 Mb and Beyond," SPIE Vol. 2197, pp. 781 – 790 (1994).

- G. Rivera and P. Canestrari, "Process Induced Wafer Distortion: Measurement and Effect on Overlay in Stepper Based Advanced Lithography," SPIE Vol. 807, pp. 806 – 813 (1993).

- H. Izawa, K. Kakai, and M. Seki, "Fully Automatic Measuring System for Submicron Lithography," SPIE Vol. 1261, pp. 470 – 481 (1990).

- K. Chivers, "A Modified Photoresist Spin Process for a Field-by-Field Alignment System," Proceedings of the Kodak Microelectronics Seminar, pp. 44 – 51 (1984).

- L. M. Manske and D. B. Graves, "Origins of Asymmetry in Spin-Cast Films Over Topography," SPIE Vol. 1463, pp. 414 – 422 (1991).

- S. Kuniyoshi, T. Terasawa, T. Kurosaki, and T. Kimura, "Contrast Improvement of Alignment Signals from Resist Coated Patterns," *J. Vac. Sci. Technol.*, B5(2), pp. 555 – 560 (1987).

- G. Flores and W. W. Flack, "Photoresist Thin-Film Effects on Alignment Process Capability," SPIE Vol. 1927, pp. 366 – 380 (1993).

- N. Bobroff, and A. Rosenbluth, "Alignment Errors from Resist Coating Topography," J. Vac. Sci. Technol. B, pp. 403 – 408 (1988).

- R. Mohondro, S. Bachman, T. Kinney, G. Meissner, and D. Peters, "High Contrast Eduction Spin Coat Process Effects on Uniformity and Overlay Registration," Proceedings of the Olin Microlithography Seminar, pp. 131 – 140 (1997).

- C. P. Kirk, "Theoretical Models for the Optical Alignment of Wafer Steppers," SPIE Vol. 772, pp. 134 – 141 (1987).

- K. Ota, N. Magome, and K. Nishi, "New Alignment Sensors for Wafer Stepper," SPIE Vol. 1463, pp. 304 – 314 (1991).

- D. L. Meunier, D. Humphrey, B. Peck, P. Feeney, J. Paquette, and J. Thibault, "Optimization of Metal Layer Overlay and Alignment Targets in a Chemical-Mechanical Polishing Environment," Proceedings of the Olin Microlithography Seminar, pp. 355 – 366 (1996).

- S-W. Hsia, G. Miyagi, and M. Brongo, "Alignment Characterization of CMP Tungsten Process," Proceedings of the Olin Microlithography Seminar, pp. 381 – 390 (1996).

- D. L. Meunier, B. Plambeck, P. Lord, and N. Knoll, "The Implementation of Coherence Probe Microscopy in a Process Using Chemical Mechanical Polishing," Proceedings of the OCG Microlithography Seminar, pp. 155 – 169 (1995).

- N. Shirishi, A. Sugaya, and D. P. Coon, "Alignment Strategies for Planarizing Technologies," SPIE Vol. 3051, pp. 836 – 845 (1997).

- T. Kanda, K. Mishima, E. Murakami, and H. Ina, "Alignment Sensor Corrections for Tool Induced Shift (TIS)," SPIE Vol. 3051, pp. 846 – 855 (1997).

- W. H. Arnold, "Overlay Simulator for Wafer Steppers," SPIE Vol. 922, pp. 94 – 105 (1988).

- B. Rangarajan, M. Templeton, L. Capodieci, R. Subramanian, and A. Scranton, "Optimal Sampling Strategies for Sub-100 nm Overlay," SPIE Vol. 3332, pp. 348 – 359 (1998).

- G. E. Flores, W. W. Flack, S. Avlakeotes, and B. Martin, "Process Control of Stepper Overlay Using Multivariate Techniques," Proceedings of the OCG Microlithography Seminar, pp. 201 – 215 (1995).

- G. E. Flores, W. W. Flack, S. Avlakeotes, and M. Merrill, "Monitoring and Diagnostic Techniques for Control of Overlay in Steppers," SPIE Vol. 2439, pp. 40 – 60 (1995).

- R. M. Silver, J. Potzick, and R. D. Larrabee, "Overlay Measurements and Standards," SPIE Vol. 3429, pp. 262 – 272 (1995).

- D. J. Coleman, P. J. Larson, A. D. Lopata, W. A. Muth, and A. Starikov, "On the Accuracy of Overlay Measurements: Tool and Mark Asymmetry Effects," SPIE Vol. 1261, pp. 139 – 161 (1990).

- A. Starikov, D. J. Coleman, P. J. Larson, A. D. Lopata, and W. A. Muth, "Accuracy of Overlay Measurements: Tool and Mark Asymmetry Effects," *Optical Engineering*, Vol. 31(6), pp. 1298 – 1310 (1992).

- M. E. Preil, B. Plambeck, Y. Uziel, H. Zhou, and M. W. Melvin, "Improving the Accuracy of Overlay Measurements through Reduction in Tool and Wafer Induced Shifts," SPIE Vol. 3050, pp. 123 – 134 (1992).

- N. Smith, G. Goelzer, M. Hanna, and P. Troccolo, "Minimizing Optical Overlay Measurement Errors," SPIE Vol. 1926, pp. 450 – 462 (1993).

- J-S. Han, H. Kim, J-L. Nam, M-S Han, S-K. Lim, S. D. Yanowitz, N. P. Smith, and A. M. C. Smout, "Effects of Illumination Wavelength on the Accuracy of Optical Overlay Metrology," SPIE Vol. 3051, pp. 417 – 425 (1997).

- Y. Tanaka, M. Kamiya, and N. Suzuki, "New Methodology of Optimizing Optical Overlay Measurement," SPIE Vol. 1926, pp. 429 – 439 (1993).

- T. Saito, H. Watanabe, and Y. Okuda, "Overlay Error of Fine Patterns by Lens Aberration using Modified Illumination," SPIE Vol. 3051, pp. 687 – 696 (1997).

YIELD:

- C. F. King, G. P, Gill, and M. J. Satterfield, "Electrical Defect Monitoring for Process Control," SPIE Vol. 1087, pp. 76 – 82 (1989).

- G. Dickerson and R. P. Wallace, "In-line Wafer Inspection using 100-Megapixel-per-Second Digital Image Processing Technology," SPIE Vol. 1464, pp. 584 – 595 (1991).

- L. H. Lin, "A Holographic Photomask Defect Inspection System," SPIE Vol. 538, pp. 110 – 116 (1985).

- S. P. Billat, "Automated Defect Detection on Pattern Wafers," *Semiconductor International*, (May 1987).

- D. L. Cavan, "Patterned Wafer Inspection Using Laser Holography and Spatial Frequency Filtering," *J. Vac. Sci. Technol.*, (November 1988).

- D. Hendricks, J. Jau, H. Dohse, A. Brodie, and D. Meisburger, "Characterization of a New Automated Electron-Beam Wafer Inspection System, SPIE Vol. 2439, pp. 174 – 183, (1995).

- J. Sturtevant, M. Chaara, R. Elliot, L. Hollifield, R. Soper, D. Stark, N. Thane, and J. Petersen, "Antireflection Coating Process Characterization and Improvement for DUV Lithography at 0.25 □m Groundrules," SPIE Vol. 2440, pp. 582 – 593 (1995).

- E. H. Bokelberg and M. E. Pariseau, "Excursion Monitoring of Photolithographic Processes," Olin Microlithography Seminar, pp. 85 – 100 (1997).

- J. R. Dralla and J. C. Hoff, "Automatic Classification of Defects in Semiconductor Devices," SPIE Vol. 1261, pp. 173 – 182 (1990).

- D. J. Friedman and S. L. Albin, "Clustered Defects in IC Fabrication: Impact on Process Control Charts," *IEEE Trans. Semiconductor Manufacturing*, Vol. 4(1), pp. 36 – 42 (1991).

- D. Michelson, "Data-Driven Method for Calculating Limits for Particle Control Charts," SPIE Vol. 2876, pp. 38 – 44 (1996).

- Y-M Chou, A. M. Polansky, and R. L. Mason, "Transforming Non-Normal Data to Normality in Statistical Process Control," *Journal of Quality Technology*, Vol. 30(2), pp. 133 – 141 (1998).

- D. Bakker and D. Icke, " Lithography Defect Characterization Using Statistical Process Control and Automated Inspection," Proceedings of the KTI Microelectronics Seminar, pp. 361 – 367 (1990).

- M. E. Clarke and K-S. Cheng, "New Photochemical Filtration Technology for Process Improvement," Olin Microlithography Seminar poster session (1997).

- Y-T. Fan, H-P. Lin, Y-C. Lo, C-H. Lee, and K-L. Lu, "A Study on Methods to Reduce Metal Defects Caused by a Coating Process," Olin Microlithography Seminar poster session (1997).

- E. H. Bokelberg, J. L. Goetz, and M. E. Pariseau, "Photocluster Defect Learning and Develop Process Optimization," Olin Microlithography Seminar, pp. 127 – 139 (1996).

- L. Galbraith and A. Neukermans, "Contamination Detection on Semiconductor Wafers," SPIE Vol. 774, pp. 13 – 20 (1987).

LINEWIDTH METROLOGY:

- K. M. Monahan, M. Davidson, Z. Grycz, R. Krieger, B. Sheumaker, and R. Zmrzli, "Low-Loss Electron Imaging and its Application to Critical Dimension Metrology," SPIE Vol. 2196, pp. 138 – 144 (1994).

- R. R. Hershey and M. B. Weller, "Nonlinearity in Scanning Electron Microscope Critical Dimension Measurements Introduced by the Edge Detection Algorithm," SPIE Vol. 1926, pp. 287 – 294 (1993).

- J. I. Goldstein, D. E. Newbury, P. Echlin, D. C. Joy, C. Fiori, and E. Lifshin, *Scanning Electron Microscopy and X-Ray Microanalysis,* 2nd edition, Plenum Press, New York (1984).

- J. Finders, K. Ronse, L. Van den Hove, V. Van Driessche, and P. Tzviatkov, "Impact of SEM Accuracy on the CD-control During Gate Patterning Process of 0.25 mm Generations," Proceedings of the Olin Microlithography Seminar, pp. 17 – 30 (1997).

- K. M. Monahan, J. P. H. Benschop, and T. A. Harris, "Charging Effects in Low-voltage SEM Metrology," SPIE Vol. 1464, pp. 2 – 9 (1991).

- L. Reimer, *Image Formation in Low-Voltage Scanning Electron Microscopy*, SPIE Vol. TT12, SPIE Press, Bellingham, WA (1993).

- M. Davidson and N. T. Sullivan, "An Investigation of the Effects of Charging in SEM Based CD Metrology," SPIE Vol. 3050, pp. 226 – 242 (1997).

- C. M. Cork, P. Canestrari, P. DeNatale, and M. Vasconi, "Near and Sub-Half Micron Geometry SEM Metrology Requirements for Good Process Control," SPIE Vol. 2439, pp. 106 – 113 (1995).

- S. R. Rogers, "New CD-SEM Technology for 0.25 mm Production," SPIE Vol. 2439, pp. 353 – 362 (1995).

- NIST SRM standards are available from the Office of Standard Reference Materials, NIST, EM 205, Gaithersburg, Md. 20899. Their phone number is (301) 975-6776. Information can be obtained from the NIST web site, www.nist.gov.

- M. T. Postek, A. E. Vladar, S. Jones, and W. J. Keery, "Report on the NIST Low Accelerating Voltage SEM Magnification Standard Interlaboratory Study," SPIE Vol. 1926, pp. 268 – 286 (1993).

- M. T. Postek, "Scanning Electron Microscope-based Metrological Electron Microscope System and New Prototype Scanning Electron Microscope Magnification Standard," *Scanning Microscopy*, 3(4), pp. 1087 – 1099 (1989).

- Y. Nakayama and K. Toyoda, "New Submicron Dimension Reference for Electron-Beam Metrology System," SPIE Vol. 2196, pp. 74 – 84 (1994).

- B. L. Newell, M. T. Postek, and J. P. van der Ziel, "Performance of the Protoype NIST SRM 2090A SEM Magnification Standard in a Low-Accelerating Voltage SEM," SPIE Vol. 2439, pp. 383 – 390 (1995).

- T. W. Reilly, "Metrology Algorithms for Machine Matching in Different CD SEM Configurations," SPIE Vol. 1673, pp. 48 – 55 (1992).

- K. Phan, J. Nistler, and B. Singh, "Metrology Issues Associated with Submicron Linewidths," SPIE Vol. 1464, pp. 424 – 437 (1991).

- E. E. Chain, M. G. Ridens, and J. P. Annand, "SPC Qualification Strategy for CD Metrology," SPIE Vol. 2876, pp. 218 – 224 (1996).

- D. A. Chernoff, "Atomic Force Microscope (AFM) Analysis of Photoresist Test Structures for use in SEM as In-House Linewidth Standards," SPIE Vol. 2439, pp. 392 – 400 (1995).

- M. T. Postek, A. E. Vladar, and M. P. Davidson, "Fourier Transform Feedback Tool for Scanning Electron Microscopes Used in Semiconductor Metrology," SPIE Vol. 3050, pp. 68 – 79 (1997).

- L. J. Zych, G. Spadini, T. F. Hassan, and B. A. Arden, "Electrical Methods for Precision Stepper Column Optimization," SPIE Vol. 633, pp. 98 – 105 (1986).

- L. W. Linholm, R. A. Allen, and M. W. Cresswell, "Microelectronic Test Structures for Feature Placement and Electrical Linewidth Metrology," in *Handbook of Critical Dimension Metrology and Process Control,* K. M. Monahan, ed., SPIE Press, Bellingham, WA (1993).

- B. J. Lin, J. A. Underhill, D. Sundling, and B. Peck, "Electrical Measurement of Submicrometer Contact Holes," SPIE Vol. 921, pp. 164 – 169 (1988).

- E. E. Chain and M. Griswold, "In-Line Electrical Probe for CD Metrology," SPIE Vol 2876, pp. 135 – 146 (1996).

- S. M. Kudva and R. W. Potter, "Cost Analysis and Risk Assessment for Metrology Applications," SPIE Vol. 1673, pp. 2 – 13 (1992).

- R. R. Hershey and R. C. Elliott, "Procedure for Evaluating Measurement System Performance: A Case Study," SPIE Vol. 2439, pp. 363 – 373 (1995).

- J. Engel and B. de Vries, "Evaluating a Well-Known Criterion for Measurement Precision," *J. Quality Technol.*, 29(4), pp. 469 – 476 (1997).

- D. C. Joy and C. S. Joy, "Low Voltage Scanning Electron Microscopy," *Micron*, Vol. 27, No. 3 – 4, pp. 247 – 263 (1996).

CONTROL OF OPERATIONS:

- *Juran's Quality Control Handbook*, 4th Edition, J. M Juran and F. M. Gryna, eds., McGraw Hill, New York (1988).

- C. A. De Angelis, "ICI Advanced Materials Implements ISO 9000 Program," *Quality Progress*, Vol. 24, pp. 49 – 51 (1991).

- H. J. Levinson, "ISO 9000: What Every Microlithographer Should Know," SPIE Vol. 2196, pp. 536 – 550 (1993).

- H. J. Levinson and C. DeHont, "Leading to Quality," *Quality Progress*, pp. 55 – 60 (1992).

- H. J. Levinson and J. Ben-Jacob, "Managing Quality Improvement on a Development Pilot Line," *Quality Management Journal*, Vol. 3(2), pp. 16 – 35 (1996).

INDEX

Harry J. Levinson has been active in lithography for nearly two decades, at companies such as IBM, Sierra Semiconductor and Advanced Micro Devices. In these positions, Dr. Levinson has applied lithography to many different technologies, including 64Mb and 256Mb DRAM development, the manufacturing of applications-specific integrated circuits, thin film heads for magnetic recording, and advanced logic. He was one of the first users of 5× steppers in Silicon Valley and was an early participant in 248 nm and 193 nm lithography. Dr. Levinson has published numerous articles on lithography science, on topics ranging from thin film optical effects and metrics for imaging, to overlay and process control. He has been an invited speaker at several lithography conferences. Dr. Levinson is also an instructor for short courses on lithography science, lithography process control, and total quality management.